Dieter

Tro
Marinelife

The identification handbook for divers and snorkellers

IMMEL
Publishing

Text © 1996: Eichler, Dieter
Photography:
DIVEMEDIA, Nürnberg: 209
P. Fischbacher: 65 bottom, 101 centre, 159 centre left, 187 bottom
E. Lieske: 195 top
A. Koffka: 89 bottom left
A. Rödiger/MTI-Press: 93 top
E. Schraml: 195 centre left
UW-Fototechnik, A. Steyr: 208
Colour plates pages 123, 145 and 171: E. Lieske

All other photographs and drawings are the work of the author.
Identification of fish and expert advice: Ewald Lieske
Identification of invertebrates: Dr Dietrich Kühlmann

Cover design: Icon Publications Ltd
Translation from German by TPS Translations
Typesetting: Shirley Kilpatrick
Printed in Hong Kong

Cataloguing in Publication Data

A CIP Record for this book is available from the British Library
ISBN 1-89-8162107

Published by:
Immel Publishing Ltd,
20 Berkeley Street,
Berkeley Square,
London W1X 5AE.

Tel: 0171 491 1799
Fax: 0171 493 5524

Contents

How to use this book

The coral reefs of the tropical seas, the most densely populated wildlife habitats on earth, are becoming increasingly popular. More and more people go there to relax by snorkelling or diving.

For many people, the pleasure of discovering something new makes them want to learn more about that discovery. The confusing variety of different marine creatures, or of forms that at first sight might hardly be taken for living things, can appear overwhelming.

This book only covers creatures that live in the Red Sea, the Indian Ocean and some adjoining areas. A guide that identified all the different species found in tropical seas would be such a bulky volume that no-one could take it on holiday, and it would also be extremely expensive. We are therefore concentrating on the identification of the most common groups of animals which can be recognised by external features. Behaviour, environment and geographical spread can provide information about the family or order. Do not expect, therefore, to see every animal you meet under the sea in the pages of this book.

The species illustrated have been identified as far as possible. The invertebrates, however, can often only be classified precisely if they are killed. As this would be unacceptable to most divers, and is also at odds with today's environmental awareness, we have in some cases dispensed with exact identification of the animals illustrated in this book. We have also dispensed with statistics of number of scales, teeth and fin rays.

The sizes given are maximum dimensions, including those for invertebrates; for these, sizes have been estimated according to the species illustrated. Every group is described on one or two double pages and illustrated with a few photographed examples. This means that you can compare visual features while reading without having to turn the page. A glossary of specialised terms is to be found on page 219.

The book contains three colour plates of drawings with descriptions of various species. These were compiled by the biologist Ewald Lieske, who has kindly allowed me to use them. Ewald Lieske has stood by me with advice during the preparation of this book, and supported me in many ways, for which I would like to thank him. I would also like to thank the biologists Dr Doris Roth, Dr Dietrich Kühlmann, Klaus Fiedler, and the director of the Sub-Aqua Diving Base Axel Horn, Ellaidhoo, Maldives, for their support. I would also like to thank the divers of the diving base SANTANA in Phuket, Thailand, who often assisted me with difficult photographs.

Biological classification

Every organism is grouped within the classification system according to its evolutionary history and relationship to other organisms. The sequence of categories within this hierarchical system is the same for each species. Here it is illustrated by the emperor angelfish (*Pomacanthus imperator*).

Sub-kingdom:	Metazoa	Metazoans
Kingdom:	Animalia	Animal
Section:	Eumetazoa	Eumetazoans
Phylum:	Chordata	Chordates
Sub-phylum:	Vertebrata	Vertebrates
Superclass:	Pisces	Fishes
Class:	Osteichthyes	Bony fishes
Order:	Perciformes	Perch-like fishes
Family:	Pomacanthidae	Angelfish
Genus:	*Pomacanthus*	True angelfish
Species:	*imperator*	Emperor angelfish

The scientific name of a species is composed of two words: the genus and the species, as for our example *Pomacanthus imperator*. In rare cases a third word can be added which covers sub-species or forms. Frequently the name of the scientist who first described the species is added.

The system is not rigid and is adjusted to take new scientific discoveries into account. For instance, only a relatively short time ago angelfish and butterfly fish were included in the Chaetodontidae family; both groups had the rank of sub-families. Today each forms a separate family.

There is, unfortunately, no uniform opinion as to the biological order of the groups. It is rather confusing for lay enthusiasts to find a different sequence in every book. The following systems have been used in this book, following expert advice:
• Invertebrates: according to Prof. Dr Rupert Riedl,
• Sharks: according to Leonard J.V. Compagno,
• Bony fishes: according to J.S. Nelson.

Although I am sure not all readers will be interested in the scientific names of the sea creatures, here is some advice which could be of help. All names of families end in -idae, sub-families (if they exist) in -inae and orders (at least in the case of fish) in -iformes. Other groups do not have any regular forms.

Identification of marine animals

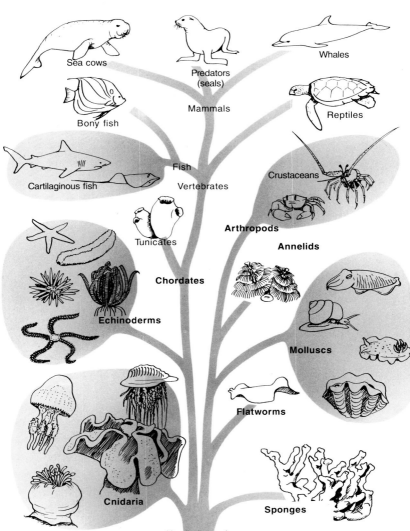

Sea cows

Predators (seals)

Whales

Mammals

Bony fish

Reptiles

Cartilaginous fish

Fish

Vertebrates

Crustaceans

Tunicates

Arthropods

Annelids

Chordates

Echinoderms

Molluscs

Flatworms

Cnidaria

Sponges

multi-celled organisms

single-celled organisms

Identification of bony fishes – Osteichthyes

13

The basics of life in the sea

Plankton –
the food of sea creatures

The term 'plankton' was first used by Victor Hensen in 1887 and means 'that which drifts along'. The word derives from ancient Greek.

Plankton consists of organisms which drift freely in the water. They are all equipped with features that prevent sinking: some can perform rhythmical swimming movements (jellyfish, shrimps), others float by means of gas or oil deposits. For others, the difference in specific gravity of the sea water and their body fluids is enough to keep them afloat.

These organisms are divided into two groups: phytoplankton and zooplankton.

Phytoplankton is plant plankton, which consists of microscopic single-celled algae, colonies of algae or larger types of algae such as seaweed. The largest floating seaweed is sargassum, which is found only in the Sargasso Sea.

Zooplankton is made up of animal organisms in various stages of development: eggs, larvae, jellyfish and fully developed animals whose body movements are not strong enough to keep them from floating with the current. Some of these animals spend all the stages of their lives in a plankton state, while others only spend particular phases of their development as plankton.

Many bony fishes spawn in open water

Their fixed way of life makes corals dependent on food brought by the currents. The main attraction of a coral reef is the variety of fascinating shapes which occur there.

Aerial photograph of an island in the North Male Atoll of the Maldives, with a few mini-atolls in the background. Without the growth of stone coral this group of islands would not exist.

and the eggs drift away as plankton. The larvae which hatch after a certain time also live as plankton, until they settle on a reef shortly before changing into fish. It is then that the young fish will begin the nektonic cycle of its life. Nekton consists of all the creatures that can move by their own power against the current of the water.

The borders between plankton and nekton are not always clear. A jellyfish which, in calm water, can move faster than the water under its own power, is nektonic; but when the movement of the water becomes stronger (for instance because of tidal currents) the same creature will be planktonic.

Unimaginable numbers of species and individuals make up the mostly minute plankton, and form the food base for many, generally small, sea creatures. These themselves become the food of larger animals. Among the plankton eaters are the corals.

What are corals?

Corals are structures usually built and inhabited by many colonies of tiny animals, the coral polyps. The materials used by the coral colonies, e.g. horn or lime, depend among other things upon the biological group of the coral. All corals, together with other

Coral reef with various types of stone coral in the Andaman Sea, Thailand

anthozoans, jellyfish and polyps, belong to the phylum Cnidaria.

The most important corals for the creation of a coral reef are:

Stone Corals

The polyps of the stone corals have the ability to secrete lime on the outer surfaces of their bodies, which they take from the sea water. When a coral polyp settles on a firm base, it first of all forms a beaker-shaped nest of lime around itself, into which it can withdraw and be relatively safe from predators (see drawing). Once the lime structure has reached a certain height the polyp reproduces by budding and more polyps develop around the primary polyp.

A further method of polyp reproduction is by splitting. One polyp becomes two, then four, then eight etc. A coral structure is not unlike a high-rise building, with the difference that the new floors are not built until the apartments on the floor below are occupied.

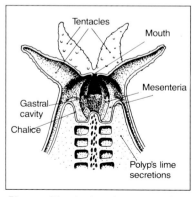

Diagram of the structure of a stone coral polyp

The many different types of polyp have different building methods, which lead to the great variety of shapes in a coral reef. The amount of work done by these tiny creatures is so immense that the largest structures built by human beings seem tiny when set against the largest coral reefs. The Great Barrier Reef off Australia's north-eastern coast is 2,000 km long!

However, not all stone corals are capable of building coral structures. There are also solitary coral polyps. The most common are the mushroom corals of the genus *Fungia*: they are like round, oval or elliptical disks and lie loose on the sea bed (see page 55 bottom).

Some species of stone coral occur at great depths and in cold waters, but these are small and not capable of building reefs.

Reef-building corals have to grow relatively quickly, which is only possible by means of one important chemical process:

There is great rivalry for space in coral reefs: here two kinds of stone coral are struggling for light and food.

Photosynthesis

The term means the creation with the aid of light of organic compounds from non-organic substances in plants. All green plants, helped by the chlorophyll which gives them their colour, produce organic matter through photosynthesis. During photosynthesis, which is powered by solar energy, water and carbon dioxide are turned into glucose and oxygen.

In the sea, too, algae produce oxygen using photosynthesis. Certain algae, the zooxanthellae, live in symbiosis within the coral polyps (see page 28). The algae partially nourish the coral with the glucose they produce, and multiply the lime secretion potential of the polyps many times. Photosynthesis, therefore, is an important chemical process for the creation of coral reefs.

$$6\,CO_2 + 6\,H_2O \;\rightarrow\; C_6H_{12}O_6 + 6\,O_2$$

$$\text{Carbon dioxide} + \text{Water} \;\rightarrow\; \text{Glucose} + \text{Oxygen}$$

Coral reef formation

Reefs can only develop where the water is sufficiently clear for photosynthesis to work. At a depth of 40-50 metres daylight has been absorbed by the filter effect of the water to such an extent that rapid coral growth is no longer possible. The second important factor in the growth of coral is water temperature. Coral grows fastest between 20 and 30°C. Lime is hardly secreted at all below 20° and above 30°C.

Due to the changes in temperature from summer to winter, reefs cannot form north of the Tropic of Cancer and south of the Tropic of Capricorn. The extent to which coral reefs thrive at the northern and southern extremities is mainly dependent on warm currents and whether or not they are constant throughout the year.

The equatorial currents which flow from east to west warm up on the surface. When in the west they meet a land mass, they separate to the north and south. This is why coral reefs off the east coasts of continents extend considerably further to the north and south than those on the west coasts.

Coral reefs either grow seawards from the shore or up from the sea bed until they are close to the surface. All the known types of reefs are formed by these two methods of growth.

The natural growth of a reef can be seen clearly in many of the atolls in the Maldives. The big lagoons of these atolls are often 25 to 50 m deep. In the shallows the currents are often stronger. This means the coral is better

Young stone coral which has built up large "foundations" to support its weight and provide resistance to the currents when it is fully grown.

Reef formation by coral growth

In shallow water with a firm sea bed coral grows faster due to stronger light and a greater food supply; a patch or hill reef results.

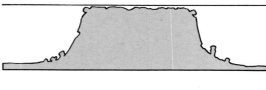

When the hill reef reaches the water surface (at low tide), it can continue growing outwards only; a platform reef results.

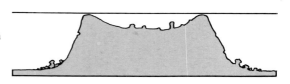

The flat of a platform reef covers a large area. Island formation due to piling up of sand is possible.

A mini-atoll occurs when the centre of a reef has sunk due to erosion and a lagoon forms. This reef type is not a true atoll.

nourished. New corals settle and reproduce.

The shallower the water, the more light reaches the coral, which speeds their growth. A patch or hill reef is formed (see drawing). When the coral reaches the surface at low tide, the reef can only extend outwards. The flat of the reef expands and a platform reef is formed. More food is available for coral at the edge of the reef, so that the reef expands continuously.

Coral reefs, however, do not just grow, but are destroyed again in certain places by natural events. A fringing reef (see page 20), which has grown almost up to the water surface, forms a kind of wall with an area of still water on one side. This reduces the current, so that the coral's plankton supply is reduced. In addition, these areas are often relatively shallow and the water temperature can rise considerably due to insufficient circulation. As we know, coral can hardly secrete lime at all above 30°C; growth stops almost completely. At night, when the water has cooled down, the solar energy for photosynthesis is missing.

The existing reef is also being destroyed from within by a great variety of organisms. Sponges, mussels, sea urchins and algae bore into the limestone by mechanical or chemical means, until coral structures or even whole parts of the reef collapse. This often happens when there are powerful storm breakers. Healthy corals can also break off during storms if they have a filigree shape. However, a reef can only be formed if the rate of growth is faster than the rate of destruction. It is the continual alternation of the two that creates the great variety of reef types.

Fringing reefs are the most common: in particular they are the dominant form in the Red Sea. As the name indicates, these reefs fringe the coast, running more or less parallel to it. The width attained by a fringing reef depends on the shape and depth of the sea bed.

The further a fringing reef extends towards the sea, the worse the conditions become for the coral between the outer reef and the land. If the destruction of the limestone has progressed far enough in this area, the sea bed will sink. A lagoon will be formed and the fringing reef will become a lagoon fringing reef. Once the lagoon fringing reef has grown far enough out to sea, it can hardly be distinguished from a barrier reef.

Even today fringing reefs are still the most effective protection against erosion of the land by the sea. No wave protection built by human hands comes anywhere near the enduring resistance of a coral reef.

Barrier reefs are distinguished from fringing reefs by the fact that they have not grown seawards from the coast, but up from the sea bed to the surface. Ribbon-like, they lie off the coastline, often many miles away from it. The most famous barrier reef is the one off the north-east coast of Australia.

Platform reefs develop from hill or patch reefs which are surrounded by a relatively even depth of water, e.g. in atoll lagoons. First of all the hill reef grows until it is just under the water surface and then extends itself in all directions to form a platform. The shape can be round or oval and it can be up to 10 km long.

Diagram of a fringing reef

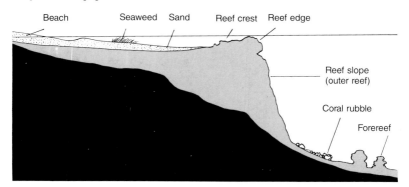

The basic forms of coral reefs
Linear reefs
Fringing reefs always run close to the coast and grow out to sea – the depth and nature of the sea bed determines the distance.

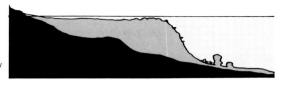

Barrier reefs lie off the coast, often many miles out to sea. They grow upwards from a maximum depth of 50 m (a fringing reef grows outwards from the coast).

Circular reefs
Platform reefs develop on a firm sea bed which must not lie deeper than 50 m; the reef flat is extensive.

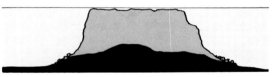

Atolls are usually surrounded by very deep water and grow in a ring shape (see also page 22, atoll formation)

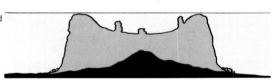

In its advanced stages the oldest part of the reef flat will sink due to erosion, forming a lagoon. Ring-shaped reefs are often know as annular reefs. If they are large enough, they are difficult to distinguish from an atoll. These pseudoatolls, which frequently occur among the atolls of the Maldives, are known as mini-atolls.

Atolls are annular reefs surrounded by very deep water, which can reach a depth of several hundred to a thousand metres.

There are many theories to account for the formation of these unusual natural features. The best known is probably Darwin's theory of atoll formation, which states that an atoll is formed when an island with a fringing reef has sunk slowly enough for the corals to grow upward in the light-filled waters near the surface. Once the island has sunk below the surface, the fringing reef becomes an atoll (see diagram page 22).

Darwin's theory has been proved in

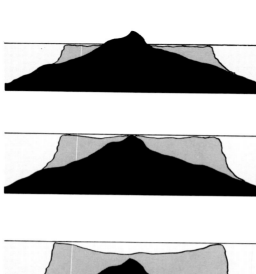

Atoll formation according to Darwin.

A fringing reef develops around a rocky island.

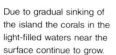

Due to gradual sinking of the island the corals in the light-filled waters near the surface continue to grow.

Once the rock has sunk below the surface, the outer corals grow faster, and a lagoon is formed.

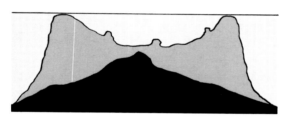

The sea bed continues to sink; only the reef ring continues to grow, and an atoll is formed.

two cases to date: in the Bikini Atoll in 1951 when drilling discovered limestone at a depth of 800 m, and in Eniwetok, where the original rock was not struck until the drilling reached 1400 m. As we know, no reef-building coral can exist in such lightless depths.

Another theory – Penck and Daly's Ice Age theory – states that the sea level was much lower in the Ice Age than it is today. At the end of the Ice Age, the coral grew upwards, following the slow rise in sea level. In the case of the two drilling operations at least, this theory cannot hold good, as the sea level would never have sunk 800, never mind 1400 metres. It is also known that temperatures on earth during the Ice Age were about 8°C lower than today. It is doubtful, therefore, whether reef-building corals could grow at that time.

However, this theory cannot be dismissed entirely, as in some regions the water temperature remains at between 26°C and 30°C all the year round. During the Ice Age, this would have been 18-22°C, which would have been enough to allow coral to grow. Also, the difference in air temperature may not have been the same for water. However, we can assume that the growth of coral reefs was only possible in relatively few areas.

The sinking of reefs or islands can occur due to their own weight on a relatively soft base, and also due to geological movements of the sea bed. Such sinking and rising is common in areas of continental movement. On an island in the Red Sea I found fossilised coral and shells, in very good condition due to the dry climate, about 30 m above sea level.

There are also fossilised coral reefs which can hardly be recognised as such – even in Europe. The limestone of the Dolomites bears witness that Europe once had a tropical climate, and that therefore coral growth and reef building were possible.

The coral reef habitat

Forms of life

Life in the sea takes on varied forms, very many of which have developed extraordinary specialisations. This is especially so around tropical coral reefs, as the pressure of competition is incredibly strong due to the high concentration of living things in a very small area. Every species that exists today has made its own so-called ecological niche, and not only in terms of physical space. Even if, for example, several species inhabit the same space, they avoid competition by making use of different foods – another possible way of occupying a niche. The density of organisms in coral reefs could only have come about because there are many possible hiding places on the reefs and the currents bring sufficient food, in the form of plankton, for so many individuals. This supply of food makes it possible for a large number of creatures to live in one place.

Coral reefs come in so many different forms that the nature of the population may vary greatly. The species living on an outer reef, for example, are quite different from those on an inner reef, and the fish living at the reef edge are different from those at a depth of 30 m. The composition of the fauna is also influenced by breakers, currents, plankton supply, algae growth and the size of the reef. Every habitat has its characteristic features and species.

On the sandy or stony bed and among the seaweed there are also specialised inhabitants which have adapted to these environments. Others live in open water, sometimes in large shoals, apparently unprotected against predators.

At night the face of the reef changes: the diurnal creatures have hidden themselves among the coral and the nocturnal creatures are looking for food – two completely different worlds?

The biology of fish

All fishes which have a bony skeleton are known as "bony fishes" and form a class of their own.

There are some 23,000 species of bony fishes; the class is therefore much richer in species than its nearest relatives, the cartilaginous fishes, although the latter are much older in evolutionary terms. While bony fishes live in waters all over the world, cartilaginous fishes mostly live in the sea; some may occasionally travel up rivers. The sturgeon is the only true freshwater cartilaginous fish.

There are about 800 species of cartilaginous fishes, divided into two sub-classes: sharks and rays. All other fish living in the sea are bony fishes (with the exception of the chimaera). The skeleton of a cartilaginous fish has no bones, and there is no bone substance in the teeth either – they are made up of dentine. Cartilaginous fishes also differ from bony fishes in having no swim bladder, no true gills or scales.

The skin of sharks is protected by tiny, mushroom-shaped denticles which are

Morphology of fish

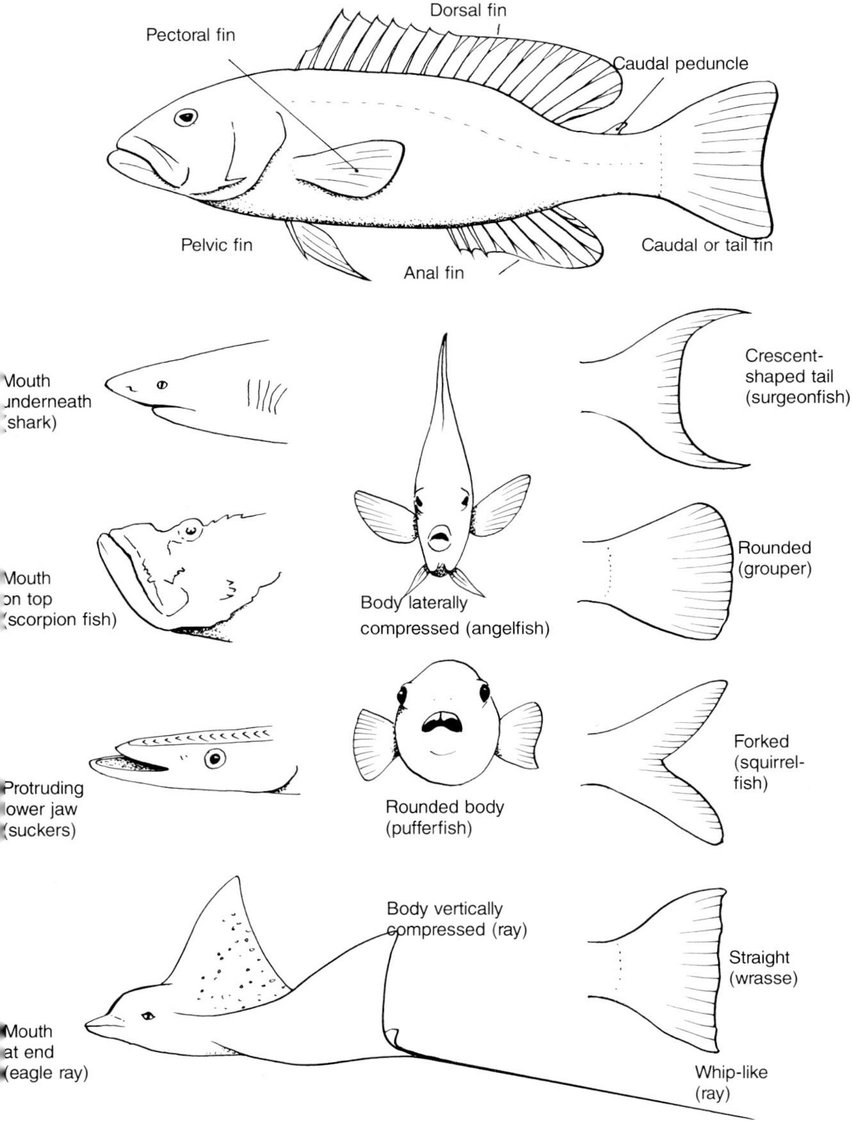

Pectoral fin

Dorsal fin

Caudal peduncle

Caudal or tail fin

Pelvic fin

Anal fin

Mouth underneath (shark)

Mouth on top (scorpion fish)

Protruding lower jaw (suckers)

Mouth at end (eagle ray)

Body laterally compressed (angelfish)

Rounded body (pufferfish)

Body vertically compressed (ray)

Crescent-shaped tail (surgeonfish)

Rounded (grouper)

Forked (squirrel-fish)

Straight (wrasse)

Whip-like (ray)

Head, body and tail forms most useful in identifying fish

very tough. It is almost impossible to cut shark skin from outside with a knife. It is so rough that it was once used as "sandpaper". All fish have paired and single fins: the pectoral and pelvic fins are paired, the dorsal, anal and tail fins are single. In many species, these are partially equipped with spines. Fish usually have good sensory organs: they can see, hear, smell, taste and feel. The development of the organs varies: predatory fish that are active by day (sharks and groupers) usually have good vision, while nocturnal predators have a good sense of smell (morays). Fish have an additional organ which land animals do not have: the lateral line. On some species it can be seen as a dotted line. Waves of pressure register through pores in the scales. This makes it possible for fish to recognise other creatures, obstacles and the nearness of the shore in conditions of poor visibility.

Fish obtain their oxygen mainly through their gills, but the skin can absorb oxygen too. Bony fishes almost all have movable gill covers, which enable the water to circulate through the gills. Cartilaginous fishes (sharks and rays) obtain their oxygen when at rest via so-called spray holes, nostril-like openings under the eyes, which can suck in water and guide it through the gills. When swimming, cartilaginous fishes open their mouths slightly, letting a larger volume of water in over the gills and thus covering their need for increased oxygen. The widespread notion that sharks need to swim from birth to death in order to obtain oxygen is only true of the deep sea sharks.

The gills of fish also have a metabolic function in that they control the

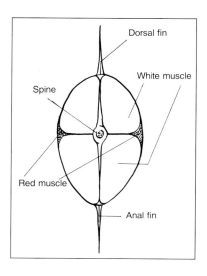

Section of a bony fish at anal fin level.

amount of salt in the blood and excrete waste products in the form of ammonia. The kidneys' main function is to produce hormones.

In contrast to mammals, fish have two types of muscle (see drawing). The red muscles work when slow, long-lasting effort is need, the white muscle is used for short, high speed activity. The white muscles with little blood supply make up most of the flesh. The red muscles with a plentiful blood supply lie along the sides between the upper and lower white muscles. They are relatively well developed in strong, fast-swimming fish such as tuna, mackerel and some kinds of shark. A fish which uses its white muscles for a long time, for instance when fighting a fishing hook, will need about 24 hours before its energy reserves have completely recovered.

For most fish growth is not hormone dependent, but continuous. The older a fish grows, the bigger it gets.

A fish by any other name

This turn of phrase perfectly describes the position as regards the naming of tropical fish. It is not just here in Europe, far away from the tropical seas, that many fish have widely differing names. Even in the tropical regions themselves, where the fish might be swimming up to your front door, the same species will sometimes have six or eight names within one country. This is not just a matter of dialect; the distance between the islands, which often lie far apart, prevents uniform naming of identical creatures.

Even the scientific names are often confused because of mistaken identification. The grey reef shark, for example, *Carcharhinus amblyrhynchos*, has also been misidentified as *C. menisorrha, C. wheeleri* and *C. spallanzani*. This shark is often erroneously termed the grey shark. There is in fact a grey shark, but it has nothing much to do with the grey reef shark. In contrast to the grey reef shark, it is extremely rare and belongs to the order Hexanchiformes. Grey sharks usually live in deep water which is not accessible to those diving for sport.

There is also a tendency in the case of English and scientific names to use the name first heard. This contributes to the continued use of names long recognised as being incorrect.

Camouflage and warning

All forms of camouflage either help peaceful fish to evade carnivores or predators to get hold of their prey. There are some remarkable specialisations in this area as well. Some species are so completely adapted in body form and colouring to their surroundings that it usually takes an experienced eye to pick out the camouflage experts. Some can change colour in seconds, adapting themselves like chameleons to their background, others imitate a shrivelled leaf drifting in the water (mimesis, see page 30). Some have the ability to bury themselves head first in the sand or to throw up sand with their fins to cover their bodies.

Fish living in open water often form large shoals for protection

Life in a shoal of fish is camouflage behaviour too. Many people believe that a predator need only swim into a shoal of fish with its mouth open and eat as much as it wants. However, fish in a shoal can easily evade such an attack. Hunters need to single out an individual victim in order to succeed. The movement of the fleeing shoal makes it easy for the predator to lose the fish it has selected in the confusion of similarly coloured bodies.

In contrast to camouflage, some fishes clearly advertise that they are dangerous, e.g. the lionfish. They are skilled hunters and have few enemies, and can therefore dispense with camouflage. Camouflage and warning can also be present in a single species. The devil scorpionfish and the spiny devilfish, which as members of the scorpionfish family have dangerous poison spines and excellent camouflage, can show warning colours by displaying their pectoral fins (see page 119).

Harmless fishes can also change their appearance in order to frighten away enemies. Porcupine and pufferfishes expand their body volume considerably by swallowing water; this makes them appear a larger adversary. Some butterflyfish change their colour and patterning at dusk, when the reef's predators go hunting. The upper halves of their sides become dark, leaving only two bright spots. In poor light the sides look like a pair of huge eyes.

Many butterflyfish have dark eye spots, usually close to the tail. The eye itself, in most species, is disguised by a dark stripe. It is assumed that this serves to confuse an attacker as to the direction in which the prey is fleeing. The butterflyfish uses this short delay to hide among the coral.

Symbiosis

Another form of protection against enemies is the partnership of two different organisms, sometimes of widely diverse species. The term symbiosis is only used when both species derive benefit from the partnership. The symbiotes can be two vertebrates, a vertebrate and an invertebrate, or an invertebrate and algae.

Symbiosis with algae is common in tropical waters. Corals, mussels and hydroids all have zooxanthellae in their tissues which have important functions (see photosynthesis, page 17).

The anemone fish, which belong to the Pomacentridae, live in close partnership with anemones, which belong to the phylum Cnidaria. The fish are tied to their anemones for life – they would hardly be able to survive without them. The only possible move is to another, nearby anemone (see also page 164).

Another well-known partnership is cleaning symbiosis. Specialised fish or crustaceans within a particular area remove the parasites on which they feed from the skin and gills of fish.

Cleaner fish station
The most common cleaner fishes, such as *Labroides dimidiatus* and *L. bicolor*, set up so-called cleaner stations. Various types of fish come to these places and indicate by signals

Porcelain crabs here *Neopetrolisthes maculatus* can often be seen in anemones – they live in symbiosis with them.

A little cleaner fish (arrow) of the wrasse family *Labroides dimidatus* is cleaning the gills of a large triggerfish *Balistoides viridescens*.

that they wish to be cleaned. Spreading the pectoral fins, the raising of the gill cover or the opening of the mouth tells the cleaner fish that this is a "client" that wants to be freed of its parasites.

Cleaner fishes do not just remove parasites from the surface of the body, they also swim into the mouth or under the gill cover. When danger threatens the cleaner fish is warned by the quivering of the jaws – a ritualised closing movement – and it leaves the mouth. In heavily populated areas one can sometimes see several "clients" patiently waiting their turn at a cleaning station.

If no client is present, the cleaner fish draws attention to itself by means of a special "dance". This conspicuous behaviour together with its colouring protect it against being eaten: cleaner fishes are taboo for predators.

The predatory false cleaner fish *Aspidontus taenatius* uses the protection accorded to cleaner fish by imitating the "genuine" article.

Mimicry

Mimicry is the copying of the form and behaviour of another species, which can serve defensive and aggressive purposes. The predatory false cleaner fish from the blenny family imitates a cleaner fish from the wrasse family in order to get close to deceived "clients". Then it will attack them and with its sharp teeth tear chunks out of the body or fins.

Mimesis

The term mimesis refers to the ability of a creature to imitate its surroundings, e.g. lifeless objects, in shape and colouring. Stonefishes and scorpionfishes are typical representatives of this group.

Squid, too, can change their relatively smooth surface in seconds so that their bodies – set with spiny-looking knobs – resemble a coral structure. This camouflage is used not only for hunting, but also when enemies appear.

A scorpionfish mimics a piece of dead coral and, well camouflaged, waits for prey

Feeding

Every organism that lives in the sea also serves as food for other forms of life. Plant matter is eaten by herbivores and the latter are eaten by predators or, if they should die a natural death, serve as food for scavengers.

The beauty of tropical reefs is often deceptive. It disguises the harsh struggle for existence that takes place there day by day: the struggle of eat or be eaten, which has been going on for two billion years. All those species that could not prevail have been pushed out in the course of evolution. Only the best adapted had a chance of survival.

Sea creatures have a very varied diet; for example, algae, kelp, seaweed, phyto- and zooplankton, bacteria, sponges, detritus, coral and fish. There is a great variety of specialised feeders. More precise information is contained in the identification section.

Invertebrates

There are many sessile forms to be found among the invertebrates. They are bound to one spot, which they usually cannot leave for the duration of their lives. They catch their food – plankton – in various ways. Sponges, tunicates and mussels are among the current feeders, which produce a current of water through their bodies by moving whip-like flagella. All the digestible matter that the creature needs for food is left hanging in the flagella.

Creatures that catch plankton by holding their feathery tentacles passively

Corals are filter feeders: they hold their tentacles passively into the current and catch plankton. The eight feathery tentacles show that this coral belongs to the Octocorallia sub-class.

into the current, e.g. tubeworms, are known as filter feeders.

Corals and hydroids are members of the order Cnidaria, which kill creatures with poison 'blisters' and hold their prey with sticky capsules. There are however species of Cnidaria which do not possess feathery tentacles, such as anemones and tube anemones.

Other sessile creatures which do not belong to the Cnidaria lay sticky traps.

These can be tentacles as well as threads of slime to which food adheres. Some snails use this method: the threads of slime are sucked in together with the particles stuck to them.

The cirripeds of the crustacean family are both current and filter feeders at the same time. They perform active strokes with their feathery legs and so catch their prey.

Among the free-moving invertebrates are the crinoids, various sea cucumbers and starfish which hold their feathery or wide-spread arms into the current. Free-moving current feeders, by moving their flagella, cause a stream of water to flow which brings the food towards their mouths.

Coral feeders do not as a rule endanger the coral reef: only the crown of thorns starfish, which feed on coral polyps, have caused great damage over the last few years. They cover stone coral and kill the polyps by secreting digestive acids: the polyps are then sucked out by the starfish and the lime skeleton of the coral is left looking white as snow (see photo on the right).

This species of starfish periodically reproduces in large numbers in some areas and may be responsible for destroying whole areas of living reef. Other coral eaters, such as certain other starfish and some types of snail do not cause lasting damage to coral.

There are also invertebrate predators which eat fish. Some species of snail, for instance, can aim and shoot poisoned darts at fish, paralysing them and swallowing them whole.

Among the grazers are sea urchins, which have a herbivorous diet and snails, which scrape up various forms of plant life with their rasping tongues.

Specialised shrimps and other crustaceans act as "cleaners" and eat major parasites off fish (see page 76).

Detritus – organic floating and sinking matter – is mainly eaten by animals living on the sea bed, such as sea cucumbers.

Fishes

There are many plankton eaters among the fishes. It is not just the smallest, but also the largest, such as mantas and whale sharks, that feed on plankton. They swim with their mouths open, often near the surface, and filter the floating organisms out of the water. Small plankton eaters, such as basslets and sleepers, swim with the current and hunt zooplankton.

Surgeonfish, blennies, rabbitfish and some species of angelfish graze on the algae on the rocks or coral stone.

Coral is eaten in a variety of ways: some specialists use their long, pointed snouts to pull the polyps wholly or partially out of their cups. Among these are butterflyfishes and some kinds of wrasse and filefishes. Parrotfishes and triggerfishes can break off branches of coral in their strong jaws or scrape away at the surface.

A large number of the creatures living in coral reefs are predators: sharks, barracudas, wrasses, scorpionfishes, snappers, morays, soldierfishes and many others. There are also omnivorous and mixed-diet fish: typical examples are unicorn and pufferfishes.

The crown of thorns starfish is an enemy of stone coral. It rapidly kills the coral polyps and sucks them out of their cups, leaving only the white lime skeleton (left, top).

The biggest fish is the whale shark. It feeds on the tiniest organisms – plankton.

Reproduction and development

Invertebrates

Reproduction in this group can be either sexual or asexual. Asexual reproduction is by division or budding.

Division means that one individual will split into two identical creatures.

Budding is the term for the growth of new individuals from a mother polyp. First of all a bud-like growth appears (hence the name), which develops into a complete animal and eventually separates off from the mother animal. In the case of stone coral, for instance, the young stay close to the mother polyp; they are identical and form a colony.

In the case of other Cnidaria, the polyp buds become medusae, jellyfish living in open water. When they reach sexual maturity they develop eggs and sperm. Larvae hatch from the fertilised eggs.

At first they live a planktonic life and then become sessile polyps. Each generation therefore takes on the form of its "grandparents", but never of its "parents". The generation change goes from medusae to sessile forms or vice versa, and occurs in almost all Cnidaria.

Apart from heterosexual reproduction, there are many examples of hermaphrodites among the invertebrates.

Hermaphrodites produce both eggs and sperm and can fertilise one another. This method of reproduction is important for maintaining the species among those creatures that rarely meet another of their kind, as it is not necessary to look for a partner of a different sex. Self-fertilisation is avoided by the sperm being ejected first, followed by the eggs.

There are two kinds of fertilisation in heterosexual reproduction -- exterior

In mating, the male cuttlefish uses one of his ten tentacles to insert a "sperm packet" into the female's mantle cavity.

and interior. Many snails have developed organs of copulation and among the Octopoda (octopus) the male uses a particular tentacle to insert a packet of sperm into the female's mantle cavity. Interior fertilisation exists among sessile species too: sponges and some stone corals whisk the sperm, which has been ejected into the open water, into their bodies when taking in oxygen and food. This fertilises the ripe eggs within the body cavities.

Some stone corals even care for their young. The larvae are not ejected into open water until they hatch. The females of many octopus species guard the eggs until the young hatch. During their watch over the eggs they do not feed and subsequently die.

In external fertilisation, sperm and eggs are ejected into open water. The fertilised eggs, therefore, develop during a planktonic stage of life.

Cartilaginous fishes

They reproduce by interior fertilisation during copulation and often bear live young (are viviparous); however, some lay eggs (oviparous). There is also a third variation in this group; the eggs, which develop in the womb, are not laid but remain in the oviduct until the young are capable of independent life and hatch (ovoviviparous).

The eggs of sharks and rays are cushion-like in shape and have spiral holding threads on the corners. With these, the eggs hang on to plants or coral and do not drift away on the current. When the young hatch they are fully developed and start to look for food.

The eggs of the sea slug (*Hexabranchus sanguineus*) form a spiral. These delicate structures can be seen in several colours.

Many species of coral reproduce sexually; male coral eject sperm into the water and female polyps whisk it into their gastral cavity.

The male and female of the viviparous sharks separate before the young are born. The pregnant females go to certain areas to give birth. This prevents the male sharks from eating the newborn young. During and after the birth, female sharks are incapable of biting and are unable to eat their young at this time when they are most vulnerable.

Bony fishes

Most bony fishes are oviparous, i.e. they lay eggs. Viviparous species are rare among the bony fishes. Generally, there is also no interior fertilisation.

The eggs of the bony fishes are usually very small, around 1 mm in diameter, and on average it takes a week for the larvae to hatch.

A larva is a creature in its juvenile phase which has a different body form from that of the adult. A comparable stage is the tadpole, a larval phase of the frog.

Fish larvae often have a yolk sac to feed from at first. Some species can feed themselves from birth on phytoplankton. They float with the current and form part of the zooplankton.

As the larvae develop, they often form spines and bony plates which provide some protection from predators.

The length of the larval stage varies considerably and can influence the spread of a species. Short-lived larvae cannot float any great distances. For this reason such species are often confined to a relatively small area. They are known as endemic species.

When the larval stage comes to an end, the young of many species have to find a suitable place to settle in relatively shallow water. They would have little chance of survival in the open sea.

Within a few days the larvae are recognisable as young fish. Pigments, scales and all fins develop in a very short space of time.

Many species of bony fishes undergo a sex change during their development. Anemone fish, for example, are born male and can if required change to female. The opposite is true of groupers, which are mostly female when young.

For many species a fish's growth is continuous, meaning that a large fish is also an old fish. Other species grow relatively quickly up to a certain size, but then more slowly. Growth can also be restricted due to social ranking, among other factors (see page 164).

Bony fishes undergo short and long term pairing, pair as couples and indulge in multiple pairing (harem). Male groupers have a harem of 6-8 females. As long as this proportion holds, no female can turn into a male. Not until the male dies does the highest ranking female change sex. The sex change is determined by the social ranking.

Bony fishes spawn in pairs or in groups. The eggs are either ejected into open water or glued to a prepared base: only a few species build nests or care for their young. Triggerfishes, anemone fishes, blennies and gobies guard and care for their eggs until the larvae hatch and then float away with the current, beginning the planktonic phase of their lives.

A special form of care for the young is carried out by the mouth incubators: one parent takes the fertilised eggs into its mouth, where they are safe from predators. Not until the eggs have developed into young fish do mouth incubators release them. With cardinal fish the male takes on the care of the young, which is probably why they have larger heads. For a short time after their release, the young fish will still slip into their father's mouth when danger threatens.

Even more unusual is the reproduction of seahorses and pipefish. The females place the eggs in incubating sacs attached to the males' belly. The eggs are fertilised and incubated in the sacs. When the young hatch, it is their father that "brings them into the world".

Some bony fishes spawn in the area in which they live, others undertake wide-ranging journeys to reach suitable spawning places. The timing of these journeys and other mating behaviour is controlled by the phases of the moon and the seasons. Fish often spawn at dusk during low tide; this means that fewer eggs are eaten by the reef inhabitants.

The survival rate of young fish is relatively low, when one considers that some of the larger fish lay up to a million eggs. Despite the numbers, only a small proportion grow up to guarantee the continuation of the species.

Many species of fish change their shape, colour and patterning during their development. This wrasse is only colourful as a juvenile, the adult is much plainer (see also parrotfishes, page 173)

Sponges – Porifera

Phylum Porifera. Some 5000 species. Size: from a few millimetres to nearly 2 m across.

Characteristics: Sponges are animals which are attached to a base and cannot move independently. Their shapes vary greatly: lumps, tubes, crusts, trees, branches, beakers, spheres and other bizarre forms. Most species have asymmetric growth and can be found in all shades of brown, grey, yellow, red, purple or blue. Patterning is rare. Pores can often be seen on the surface.

Found: World-wide, plentiful in depths of up to 50 m. Mainly attached to stones or rock, but also to coral and the shells of molluscs, under overhangs and in caves.

Way of life: The surface of a sponge is a layer of cells perforated with pores. Inside lies a central hollow space fitted with flagellum-bearing cells which is linked to the pores by narrow channels. By means of their whiplash motion the flagella create a current which brings oxygen and food to the sponge. Particles of food adhere to the flagella. The filtered water is forced through a large exit, the osculum. A sponge can have several oscula. This form of feeding means the sponges are independent of sea currents, which is why they are often found in shady relatively unpopulated places, such as caves and under overhangs. In the inner cell layers, skeletal features, spicules, are made of lime (calcareous sponges) or silica (glass sponges). The spines can be bound with elastic spongin to form a structure. The hard lime or silica spines which protect the animal against being eaten can cause skin irritation in humans. Sponges do not have muscles, organs or nerves. If a sponge is damaged, for instance by an anchor, the pieces can regenerate themselves under favourable conditions.

Food: Plankton and floating material such as bacteria, small algae, detritus.

Reproduction: Mainly asexual via budding, but also sexual. The sperm ejected by male sponges into the water will reach a female, and fertilisation occurs. The development of the egg to the larval stage takes place within the mother. The larvae then begin their planktonic stage.

Branching sponge: about 30 cm. The shape of the growth is reminiscent of coral. Photo: Maldives.

Branching tubular sponge, *Siphonochalina* sp.; about 60 cm. This species is common in the Red Sea. Photo: Egypt.

Red Ball Sponge, *Cinachyra* sp.; about 15 cm. Photo: Maldives.

Fire Corals – Milleporidae

Family Milleporidae, Class Hydrozoa, Phlyum Cnidaria. (See also page 42.) Size: about 1 m.

Characteristics: Very similar to stone coral, but can be distinguished by their beige to brown colouring with white thickened tips. They grow in fan shapes, in a grid or antler pattern, sometimes also forming flat surfaces. They usually face in to the main current.

Found: In tropical seas, in relatively shallow water.

Way of life: Fire corals are diurnal, as the feeding polyps are surrounded by defence polyps which give off a strong irritant. They serve not only as a defence against enemies but also catch prey, which they pass on to the feeding polyps. Fire corals secrete lime, and are therefore capable of forming structures. The porous lime skeleton surrounding the bodies of the polyps has many channels (stolonae) running through it, which link the polyps and provide them with food. They live in symbiosis with zooxanthellae.

Food: Zooplankton.

Reproduction: Fire corals develop specialised polyps from which tiny planktonic medusae break off.

Warning: The poison secreted by all cnidarians can cause painful burn-like irritation.

Hydroids – Hydroidea

Order Hydroidea, Class Hydrozoa. Some 2000 species. Size: up to 15 cm.

Characteristics: The hydropolyps form colonies shaped like feathers, slender trees or bushes. They often settle on a firm base. Colours vary.

Found: World-wide, in shady places.

Way of life: Because of their sedentary life they depend on currents which bring them food. The individual polyps are surrounded by beaker-shaped protective casings, into which they can retreat.

Food: Mainly animal plankton.

Reproduction: The hydroid colony grows by means of asexual budding. Specialised buds produce hydromedusae, little jellyfish-like forms, which swim in the water by pulsating. The hydromedusae reproduce sexually, i.e. an egg is fertilised and then develops into a larvae. When the larva settles on the bottom, another hydropolyp is formed, which forms a new colony by budding. One species, therefore, exists in two totally different forms.

Branching Fire Coral, *Millepora tortuosa*; up to 2 m. They settle on raised ground close to the surface. Photo: Thailand.

Plank-like Fire Coral, *Millepora platyphylla*; about 1.5 cm. All fire corals have stinging capsules. Photo: Thailand.

White Stinging Hydroid, *Lytocarpus philippinus*; about 10 cm. These plant-like colonies of animals have a powerful sting. Photo: Thailand.

Black corals – Antipathidae

Family Antipathidae, Order Antipatharia, Sub-class Hexacorallia, Class Anthozoa, Phylum Cnidaria. Several metres tall at depths of 50 m. About 150 species.

Characteristics: Colony-forming, tree-like corals with strong trunk. Thin, reddish-brown horny spines, reminiscent of pine needles and appearing dark green under water, are to be found at the ends of the branches. The horn-like hard skeleton is dark brown or black and fixed to a firm base by an adhesive plate. Jewellery and rosaries are made of this coral.

Found: In tropical waters, usually at depths below 10 m.

Way of life: Black coral grows very slowly and takes many years to form a large "tree". The largest ones grow in current-exposed locations at depths of about 50 m. These beautiful colonies of creatures, however, often fall victim to human greed.

Food: Microplankton and detritus.

Reproduction: Polyps reproduce sexually: male and female polyps can be present in one coral. Asexual reproduction is by budding, which increases the size of the colony.

Bushy black coral, *Antipathes* sp.; about 1.8 m. These colonies of creatures can reach several metres in height. Photo: Thailand.

Whip Corals – Cirripathes

Genus Cirripathes, Family Antipathidae. Up to a maximum length of 6 m.

Characteristics: Long, slender corals which hardly taper at all as they grow. They are never branched and have no universal form. They do not grow in straight lines and rarely grow upwards, growing instead from the reef outwards towards open water. The irregular kinks often follow a spiral path. When looked at closely, the coral polyps can be seen. Coloration is brownish.

Found: In tropical seas; usually below 15 m on a firm base; most often on steep reefs with strong currents.

Way of life: Whip corals have a strong, tough, elastic skeleton, which cannot be pushed more than slightly out of place even by a strong current.

Black whip coral, *Cirripathes* sp.; about 1 m. Photo: Maldives.

Warning: In the vicinity of this coral, take care not to get entangled with it, for instance by any slits in your flippers. This coral will neither break nor be cut.

Tube Anemones – Cerriantharia

Order Ceriantharia, Sub-class Hexacorallia, Class Anthozoa.
Size: up to 40 cm.

Characteristics: The elongated body of the ceriantharian
lives in a self-made tube, which extends into the sea bed
sediment. The tube consists of a gelatinous substance which
holds the sand together. When danger threatens the tube
anemone quickly withdraws into this tube. The tube can be
up to 1 m long and often ends around sea bed level, so that
the outer ring of tentacles lies on the bottom as a sticky trap.
The pointed tentacles have many adhesive capsules.
However, the tube can end well above sea bed level. Typical
features are the much longer tentacles of the outer ring and
the short, slightly fluorescent tentacles of the centre. Colours
vary from brown, violet, yellow, white to green.

Found: In tropical and cold waters, at depths of up to 35m,
on sea bed with sediment.

Way of life: The sedentary tube anemone has only limited
movement within its tube. It has no skeleton and, like all
anthozoans, catches prey that floats past.

Food: Zooplankton and other small creatures.

Reproduction: Sexual. Larvae are planktonic.

Daisy coral – Zoantharia

Order Zoantharia, Sub-class Hexacorallia, Class Anthozoa,
Phylum Cnidaria. About 10 species, only a few centimetres
high.

Characteristics: Polyps, almost without exception colony-
forming, which build a pipe-like outer casing out of the
silica particles of sponges and single-celled creatures. The
polyps often stay retracted within the pipe, but come out at
night and when the current is strong to catch their prey.
They usually have 24 or more yellow tentacles, surrounding
an elongated yellow mouth. The polyps are linked to one
another by channels (stolonae). When the polyps have
withdrawn, the jagged upper edges of the skeletons can be
seen.

Found: World-wide, in current-exposed locations, in caves
and steep cliffs, often in large numbers.

Way of life: Their stinging poison protects them from most
fish and grazing invertebrates. Their sedentary life makes
them dependent on food brought by the current.

Reproduction: Probably oviparous hermaphrodites.

Sea Anemones – Actiniaria

Order Acitiniaria, Sub-class Hexacorallia. Thirty-five species. Size: up to 1.5 m, making these the largest polyps.

Characteristics: The body, usually disk-like, is wide open by day, so that the many tentacles can be seen. Anemones often live close together with several others of the same species and form a kind of "lawn". When it gets dark, the tentacles disappear into a balloon-like mantle. The individual animals can then be easily distinguished. The large "foot" at the base can also be seen clearly – the creature uses this to attach itself with adhesive capsules to a firm base. In food-rich waters it may happen that they close while it is still day, probably because they have had enough food. Anemones have tentacles of varying lengths. Their colouring is unobtrusive, grey-green or beige, in rare cases fluorescent and reddish. The mantle is often brightly coloured: violet, blue or red, as well as grey, brown and greenish. In tropical waters anemone fish can often be seen among their tentacles; however, these only live in symbiosis with 10 species of anemone.

Found: In warm and cold waters, usually in relatively shallow water; the little parasite anemones live on the shells of hermit crabs.

Way of life: Anemones lead a sedentary (sessile) life, but can move by crawling slowly. They have no skeleton. Symbiosis often occurs with anemones. The best known is with anemone fish. In between the stinging and adhesive tentacles the anemone fish is safe from predators. For its part, it defends the anemone against specialised enemies which try to bite off its tentacles. It also brings large pieces of food, which the anemone cannot reach, from its surroundings. Various species of crustacean can also be seen in anemones; they are obviously also immune to the stinging toxin. The little parasite anemones, which live on the shells of hermit crabs, are carried about and have access to more varied forms of food; the crab is protected by the stinging tentacles (see also photo page 81).

Food: Zooplankton and fish.

Reproduction: Usually sexual. Some species care for the eggs until the larvae hatch. Others reproduce asexually by splitting at the base or the division of the whole animal. They have no medusa stage.

Sea anemone, *Heteractis magnitica*; about 80 cm. By day usually only the tentacles are visible. Photo: Egypt.

Common sea anemone, *Heteractis magnitica*; about 40 cm. At twilight the mantle contracts until the tentacles are covered. Photo: Maldives.

Sea anemone, *Heteractis magnitica*; about 20 cm. Photo: Thailand.

Stone Corals – Madreporaria

Order Madreporaria, Sub-class Hexacorallia, Class Anthozoa, Phylum Cnidaria. Some 2500 species living today. Size of polyps: 1 mm to c. 40 cm. Size of corals: up to several metres.

Characteristics: Stone corals, with a few exceptions, are hard, immovable, firmly fixed structures. There is a great variety of forms. Balls, pillars, beakers, plates, trees and antlers are among the shapes that can be found. Stone corals are mainly brownish or greenish in colour. A colony is built up and inhabited by many very small coral polyps. The polyps can or cannot be seen according to species. By day the creatures withdraw, but at night they stretch out their tentacles to catch food. Every polyp has six non-feathery tentacles. In the centre is a small, elongated mouth opening.

Found: World-wide in tropical seas, at depths of up to 50 m. In sunlit waters with a minimum temperature of 18°C throughout the year.

Way of life: Coral polyps produce lime (calcium carbonate) to build their ever-growing skeletons. It is easily available from seawater in the form of calcium. The chemical change from lime to limestone is a simple process in the atmosphere, achieved by taking up carbon dioxide (CO_2). Underwater, however, it is a very complicated process. A coral reef can only be created when the build-up of the lime skeletons proceeds faster than their destruction. Reefs can be destroyed by many factors; we will return to those later. Coral polyps can only produce lime fast enough to build reefs with the aid of zooxanthellae – tiny algae. The polyps live in symbiosis with the algae: the latter are present in large numbers in the tissues of reef-forming (hermatypical) stone corals. Zooxanthellae can build up organic compounds by photosynthesis and during this process take up carbon dioxide from the body of the polyp. This promotes the secretion of aragonite crystals, the building blocks of the coral's lime skeleton. Lime production increases tenfold. Reef-building corals grow more slowly the deeper they are, as they receive less light. Most species of stone coral exist only at depths above 50 m. There are also scattered corals in deep water, where no ray of sun ever penetrates. Without the symbiosis of zooxanthellae and coral polyps there would be no Great Barrier Reef in Australia, and no Maldives Islands.

Staghorn coral, *Acropora elseyi*. This species and similar relatives often colonise large areas. Photo: Maldives.

Staghorn coral, *Tubastrea coccinea*; about 6 cm. By day the polyps withdraw. Photo: Maldives.

Stone Corals (continued)

Disk coral, *Turbinaria frondens*; about 70 cm. Photo: Maldives.

Although the polyps are limited almost completely to daytime for producing coral, they are nocturnal animals which emerge from their lime fortress in darkness, when the coral fishes are sleeping. If they were to stretch out their tentacles for plankton during the day, they would soon fall prey to many coral fish species – such as some of the file and butterflyfishes. As soon as an animal plankton organism makes contact with a polyp, it is stunned or killed by the stinging capsules and brought towards the mouth opening. The poison-bearing stinging capsules, which are concentrated in large numbers particularly on the tentacles, also serve as a defence against enemies. Corals are, however, not entirely dependent on this method of feeding.

Stone corals grow most rapidly in food-poor waters at a temperature of 25-30°C. This is because the light, which is so important for photosynthesis, can penetrate deeper into the water due to the lesser amount of plankton. Because of the poor food supply in the water corals have to find additional ways of feeding. Nature has equipped them with microscopically fine cilia, which are found on the upper surface of the body. All particles which are recognised as food are transported to the mouth opening. Indigestible matter, such as sand, is pushed out. In this way the coral cleans itself. Even microscopically tiny organisms, such as bacteria, are used as food. Polyps can cover their surfaces with slime to which the micro-organisms stick. The cilia then move the slime towards the mouth. Polyps also absorb sugars and amino acids – the metabolic products of the zooxanthellae. The final, though minimal, source of nourishment comes from organic solutions which are found in sea water and can be absorbed through the body surface.

Finely branched table coral, *Acropora spicifera*; about 1.5 m. These colonies of creatures can grow up to several metres across in favourable circumstances. Photo: Thailand.

Warmth is also very important for the chemical process. This is why there are no coral reefs in areas where cold water rises from deeper levels. A further condition for life for the polyps is the salt content of the water. When the coral lies dry in the hot sun at extreme low tide, they can survive for 1 or 2 hours due to a protective layer of slime. However, if it rains during this period, the polyps will die in a very short time. In no part of the world is coral to be found near the mouths of rivers. This is due not only to the lower salt content of brackish water but also because all rivers carry huge amounts of sediment, which is the greatest enemy of coral, down into the sea.

Soft Corals – Alcyonacea

Order Alcyonacea, Sub-class Octocorallia, Class Anthozoa, Phylum Cnidaria. Size of colonies: up to c. 1m.

Characteristics: Often tree-like, fleshy colonies with thick stalks. The branches, covered with polyps, can look like flowers (Anthozoa = flower-like animals); stalk and branches are often transparent, so that the spikes of lime (sclerites) they contain can be clearly seen. In some species the sclerites stick out of the branches like thorns in the vicinity of the polyps. The genus *Sarcophyton* has a mushroom-like form, but the rim of the cap is very corrugated, which increases the surface area of the colony. These species are pale grey. The active motion of the Xeniidae family is unusual; their polyp arms can open and close rhythmically like blossoms. Currents and eddies are created by these movements, and a greater volume of water is filtered. The polyps of soft corals are relatively large and are positioned on long stalks; they are white, pale grey or beige. Soft corals are also know as leather corals. The name probably comes from the genus *Sinularia*. Its species cover dead coral structures with a thick leather-like layer with raised bumps and ridges. *Sinularia* is a strong competitor for space of stone coral. Colour: beige or greyish blue. Apart from the Xeniidae, all soft corals can retract into their structures.

Found: In the Red Sea, Andaman Islands and Nicobar Islands. They are however also to be found in other areas, some species even live in cold waters.

Way of life: Soft corals are tube-like structures which remain stable due to water pressure within. This is controlled in a valve-like manner by specialised polyps. At night the soft corals become more active and reach their full size due to increased water pressure. The lime sclerites are a further support. Soft corals have no inner skeletal structure. They settle on hard ground, such as rock or coral stone, but also on shells. The polyps are connected by means of tubes which enable them to pass food on to one another. If this "social welfare programme" did not exist, some parts of the colony would receive no food, the internal pressure could not be maintained, and the coral would sink to the bottom.

Food: Zooplankton.

Reproduction: Sexual. The larvae live as plankton, otherwise little is known.

Soft coral, *Nepthya* sp. at night; about 70 cm. At night the creatures are more active and reach their full size due to increased internal pressure. Photo: Thailand.

Soft coral, *Dentronephthya* sp.; about 60 cm. By day they often let their "branches" droop. Photo: Egypt

Soft coral, *Sarcophyton* sp.; about 80 cm. The white dots are the polyps. Photo: Maldives.

Horny Corals – Gorgonacea

Order Gorgonacea, Sub-class Octocorallia, Class Anthozoa, Phylum Cnidaria. Some 1200 species. Size of colonies: up to 3 m.

Characteristics: Judging by their shape they could be taken for plants. They grow in two or three-dimensional forms and are often strongly branched. The individual branches are relatively thin and covered by many polyps, often close together. Two-dimensional species orient facing into the prevalent current. The whole fan often grows from a short stalk, which is attached to a firm base. The adherent surface is frequently either larger than the stalk in surface area or divides like roots. The skeletons of the colonies consist of an elastic horny substance, gorgonin, which is strengthened by lime deposits (sclerites). In areas where they occur frequently the sclerites play a small part, after the death of the coral, in the formation of reefs. They can be found in the greatest variety of colours: red, orange, yellow, violet; very large specimens are usually purplish grey; the polyps are white or yellow. Belonging to the order of horny corals are the whip corals, which usually grow upwards straight from the bottom. They taper towards the tip, which often hangs down like a whip. They are not branched and are glowing red or beige in colour. All species have eight-rayed, feathery polyps with a lovely symmetry.

Found: In tropical to cold seas, especially common in the West Atlantic. In the Red Sea and the Pacific.

Way of life: Horny corals, due to their sedentary life, depend on the drift of plankton. Their fan shapes, which often have a large surface area, are ideal for filtering plankton from the water. However, many larvae also settle on them, which in part overgrow the coral. Among their enemies are various species of algae, sponges, tube worms and mussels. Horny corals are surrounded by a very delicate layer of living tissue, which can easily be damaged by mechanical movement. The layer consists of polyps and a supporting tissue containing digestive channels. The polyps are linked to one another by these channels, which secure the food supply for the whole colony. If, through certain circumstances, part of the colony cannot catch any plankton, it is fed by other polyps.

Food: Plankton.

Reproduction: Heterosexual: eggs and sperm are contained in the gastral cavity of the polyps. The eggs are fertilised by sperm drifting in and they are then ejected into the open water.

Gorgonian fan coral, *Supergorgia* sp. Horny corals have a delicate, living layer of tissue, which can easily be damaged if touched. Photo: Egypt.

Whip coral, *Juncella* sp.; about 80 cm. The polyps can be seen even by day. Photo: Egypt.

Turbellarians – Turbellaria

Class Turbellaria, Phylum Platyhelminthes. Some 2500 species. Size: up to 6 cm long.

Characteristics: Observing these delicate creatures, it is hard to believe that they are worms. They are reminiscent of slugs, but the latter mostly have feathery gills, warts, ridges or hump-like bulges on their backs, which the turbellarians do not. Their extremely compressed bodies consist of a broad foot which becomes a thin fin-like fringe at the edges. There are two feelers at the front end of the worm. In some species the fringe forms two ear-like funnels, lying relatively close together. The eyes, darkly pigmented, lie behind the feelers; some species have four eyes, others none. The upper surface of the body is thickly covered with cilia. Many species are brightly coloured and conspicuously patterned.

Found: In all waters, whatever the sea bed, also on coral. They occasionally swim close to the bottom.

Way of life: Turbellarians have an unusual way of moving. Circling eddies of water are created around the body by the movement of the cilia. Also, the long fibres of the crawling muscles are rhythmically contracted in waves from front to rear, thus pushing the body along on its slimy secretions. This combination gives the creature a harmonious motion such as is rarely seen. The elegant swimming motion of turbellarians is hardly less spectacular than their crawl. The creatures are diurnal and nocturnal. Their regenerative ability replaces missing body parts and whole new animals can grow from discarded parts. Not only herbivores belong to this group, there are also predators. Despite their slow movements they can catch other creatures. Having tracked their prey, they crawl up to it slowly and pull their bodies together. In the moment of attack, the front end is suddenly extended and the mouth opening, which can be everted like a trunk, grabs the prey.

Food: Algae, floating and sinking matter; predators, small invertebrates.

Reproduction: All flatworms are hermaphrodites which fertilise each other by copulation. The little round eggs are laid individually, but there are stalked clutches with 30-40 eggs. The larvae rarely live as plankton, they mostly develop on the bottom. The larval stage lasts 1-3 weeks.

Tiger flatworm, Order Polycladida; about 6 cm. Turbellarians can be elegant swimmers. Photo: Djibouti.

Flatworm, Order Polycladida, brightly coloured; about 5 cm. The stinging poison of the stone coral over which it is crawling is obviously harmless to this worm. Photo: Thailand.

Prosobranch Gastropods – Prosobranchia

Sub-class Prosobranchia, Class Gastropoda, Sub-phylum Conchifera, Phylum Mollusca. Some 20,000 species living in the sea. Size: up to 60 cm long.

Characteristics: The body is divided into head, muscular foot, and mantle, which are contained in an asymmetrical spiral shell. Marine gastropods differ from land snails only by their gills, which are contained in the mantle cavity together with other inner organs. The head of the gastropod is equipped with antennae, sometimes with eyes and have a file-like tongue. The outer skin of the gastropod mantle has the ability to secrete lime and in this way creates the shell. When endangered, the whole creature can withdraw into its shell and can cover the opening with a fitted cover, the operculum. The shells differ not only in size, but also in shape, colouring and patterning. Most species have flourishing growths on their shells, which help to camouflage them in their natural surroundings. Only the family Cypraeacea, to which the famous cowry shells belong, draw their mantles over their shells. This means that, while they are alive, no growths can form on their shells.

Found: Widespread in all waters, from the tidal zone to the depths.

Way of life: Gastropods move forwards creeping slowly on their foot. At the same time great quantities of slime are secreted. Not all gastropods are harmless grazers; many species are predatory. They attack other shelled animals and bore through the shells of their victims with their sharp file-like tongues. There are even some gastropods which are able to catch fish. They can aim and fire poison darts, paralysing the fish. The predatory gastropods have eyes on stalks, which can look out of the shell even when the animal has withdrawn.

Food: According to species, e.g. plants, plankton, molluscs, hydroids, sponges.

Reproduction: Usually heterosexual. Fertilisation is often by copulation but also by the release of sperm into the water. Relatively long larval stage. Many larvae have a "sail" which becomes detached when they begin the benthic (on the sea bed) stage of their lives.

Warning: Some varieties can shoot poisoned darts which can also be dangerous to humans.

Triton's trumpet, *Charonia tritonis*; about 35 cm. Many divers pick up gastropods and let them fall again, frequently leaving the soft animal exposed to predators until it can turn over and crawl away. Photo: Egypt.

Marine snail, *Tectus dentatus*; about 8 cm. These snails are well camouflaged by the growths on their shells. Photo: Egypt.

Egg cowry, *Ovula ovum*: about 7 cm. Egg cowries can totally cover their shells with their mantle. This means the shells are always smooth. Photo: Thailand.

Opisthobranch Gastropods – Opisthobranchia.

Size: up to 30 cm long, but mostly around 5 cm. About 4500 species living in the sea.

Characteristics: The opisthobranch gastropods, commonly known as sea slugs or nudibranchs, have no shell (except for a few species). Their gills are placed behind the heart. The body generally consists of the head (with feelers), the muscular foot and the mantle. The gills are on the rear half of the body. In some species the gills are missing altogether. Oxygen is then absorbed through the body surface, which is increased by humps, warts and ridges. The body form is extremely variable, ranging from oval to elongated and thin. Nudibranchs often have fringing fins or a great variety of extrusions. Species belonging to the elongated sea slug family have a very thin, elongated body with many attachments, mostly along the sides, which are ordered symmetrically. Nearly all species are splendidly colourful and conspicuously patterned.

Found: In tropical to cooler seas, on all types of sea bed, most frequent in water that is not too deep.

Way of life: Nudibranchs move forward by crawling, like their relatives the shelled gastropods. In doing so they secrete slime from glands at the front of the foot. Many kinds can also swim. When swimming the fringing fins move in large, wave-like swoops. These elegant, graceful movements are fascinating to observe. Especially large species like the Spanish dancer, which is coloured red and white, move impressively. The unshelled, apparently unprotected creatures can secrete through their outer skin a substance which frightens off enemies. There has been no evidence of poison, but secretion must be effective, as the gastropods are not attacked despite their conspicuous appearance. The contrasting colours probably act as a warning, since these gastropods do not hide.

Food: Organic sediment, seaweed: also meat.

Reproduction: Almost without exception nudibranchs are hermaphrodites. Mating has been observed in chains, with the first animal taking on the female, the last the male role. All those in-between took on both roles. The eggs, in covers of slime, are laid in strings or ribbons, often in a spiral shape. The length of the planktonic larval stage varies and covers 1-2 stages.

Pyjama nudibranch, *Chromodoris magnifica*; about 6 cm. These unprotected gastropods are not poisonous, are conspicuous, and yet are not attacked. Photo: Philippines.

Nudibranch, *Chromodoris tertianus*; about 6 cm. The gills on the backs of the opisthobranchs are positioned behind the heart. Photo: Maldives.

Bivalves – Bivalvia

Class Bivalvia, Phylum Mollusca. Some 8000 species. Size: 0.5-60 cm, larger in rare cases.

Characteristics: The body is enclosed in two shells. The two halves of the shell are sometimes symmetrical, sometimes asymmetrical; in some groups one shell can be firmly fixed to the ground. Both halves are connected by a band of tissue, the ligament, which is elastic. They are held together by powerful closing muscles. The shape of the shells varies from round to elliptical, from wedge-shaped to an elongated sheath. The shells are often splendid in colour and pattern, but are frequently covered with unattractive growths on the outside. Often the creatures are attached to the sea bed with threads (byssus filaments) which consist of dried adhesive secretions. The shells, when open for feeding, often allow a glimpse of the edge of the mantle, which in a few species can be everted. Mantles are often splendidly coloured and patterned in iridescent shades. Some species have a large number of simple sensory organs on the edges of the mantle. These register movement in the vicinity and lead to the rapid closure of the shells.

Found: In all oceans, from the tidal zone to the deep sea, on all types of sea bed.

Way of life: Only a small proportion of the many species of bivalves can be seen fastened to the sea bed or to rocks. The larger number by far live in hiding; buried in sand or mud, but also in wood or rock, into which they can drill mechanically or by means of lime-dissolving secretions. They also drill into coral rock till corals or even whole sections of reef break off or collapse. Some species stop their destructive activity when they are securely bedded into the rock, keeping only their breathing duct free, through which they also waft in their food. Only a few bivalves can move at any speed. Cockles can jump, using their bent-over foot, and some scallops can swim by jet propulsion, opening and closing their shells. Giant clams can live for more than 100 years. Enemies of the bivalves are the larger starfish, crustaceans, predatory gastropods and fish. Cases of symbiosis with small crustaceans are also recorded.

Food: Plankton and drifting matter are swept in and filtered by net-like gills. Bivalves can also store toxic substances in their bodies.

Reproduction: Eggs and sperm are ejected into open water.

Thorny oyster, *Spondylus varius*; about 30 cm. Bivalves have simple sensory organs on the rim of the mantle. These can sense enemies. Photo: Maldives.

Giant clam, *Tridacna squamosa*; about 80 cm. Some species of bivalve can evert their mantle flaps. Photo: Maldives.

Cuttlefish – Sepioidoea, Squid – Teuthoidea

Orders Sepioidoea and Teuthoidea, Class Cephalopoda, Phylum Mollusca. The class contains 650 species. Size of the largest measured giant squid (deep sea creature): 21.95 m.

Characteristics: The slightly flattened body consists of head and mantle with mantle cavity. The mantle has a rounded, sack-like form and is partially or completely bordered by a fin. There are 10 tentacles on the head, arranged around the beak-like mouth. Two of them are greatly elongated for catching prey, and can be withdrawn into pockets. The eight short arms are kept close together when swimming and extend the head out to a point. There are four rows of suckers on each tentacle. The catching tentacles are very slender and have a large number of suckers on the leaf-shaped broadened front end only. Cephalopods have very highly developed large eyes, which are similar in their structure to the eyes of vertebrates.

Found: World-wide, from close to the surface to depths of 5000m.

Way of life: Squid are very good swimmers and can swim backwards as well as forwards. They move their fins in a wave-like motion while swimming. The jet propulsion effect of the water ejected while breathing assists movement. Squid prefer to live in the open sea, but come to reefs in search of food and to mate. Cuttlefish prefer the waters above sandy bottoms and like to dig themselves in during the day. They are active at night and in twilight. All cuttlefish and squid can change colour extremely quickly for camouflage purposes. However, the "ink" which is produced in glandular cells and stored in the ink sac is just as effective a protection. When predators attack their supposed prey, the creature takes flight with a powerful backward thrust and ejects dark secretions, which contain melanin. This can have a paralysing effect on the sense of smell of some predators. Cuttlefish and squid also change colour and become pale, almost colourless, and change direction. The predator attacks the ghost image – the cloud of ink – as it cannot see its prey any more.

Food: Fish, prawns and marine snails.

Reproduction: Cuttlefish and squid gather to mate near shallow reefs or in coastal waters. Squid come in swarms and swim in pairs to corals under which they stick their eggs. The creatures embrace while mating. The male uses a special tentacle to introduce a packet of sperm into the female's mantle cavity. During mating the fringing fin has a conspicuous bright blue streak.

Cuttlefish, *Sepia* sp.; about 40 cm. During mating cuttlefish acquire a bright blue strip at the base of the fringing fin. Photo: Thailand.

Squid, *Sepioteuthis lessoniana*; about 25 cm. When spawning they swim in pairs to a suitable hiding place; while the female sticks down the eggs, the male waits outside. Other couples wait patiently until they can go into the cave. Photo Thailand.

Octopus – Octopoda

Octopus, *Octopus macropus*; about 70 cm. Not common in tropical waters. Photo: Egypt.

Order Octopoda, Class Cephalopoda, Phylum Mollusca. Maximum length is rarely more than 2.5 m. Some 650 species.

Characteristics: The body consists of a large, sack-shaped mantle and a head which has eight more or less equal length tentacles. When fully grown, an octopus can have almost 200 suckers. In the midst of the tentacles is the mouth with its powerful jaws. The poisonous bite of this "parrot beak" kills the prey. The highly developed, lensed eyes are set in protrusions at the highest point of the head. The octopus has no hard plate or skeleton. Even large animals can squeeze through the narrowest gaps. For camouflage purposes it can not only change its colour very quickly, but can even change the surface of its skin to blend with its surroundings. When mating or otherwise excited the octopus' colour deepens intensely. These changes are controlled by the many colour-bearing chromatophores which can increase in size up to 20 times. Octopus probably have well developed senses and due to their highly developed nerve systems are probably the most intelligent invertebrates.

Found: World-wide: it prefers a firm sea bed.

Way of life: Octopus move along the sea bed with a crawling motion, propelled by their arms. When danger threatens they try to hide in a cave or cleft. If none is available, they take flight by forcing water out of their bodies and propelling themselves sharply backwards. In this way they can swim quite fast, but not for very long.

Food: Fish, crustaceans, mussels.

Reproduction: The males have a specialised tentacle with which they introduce spermatophores into the female's mantle cavity. The numerous eggs (up to 100,000) are usually attached to the roofs of caves or under stones by the female, who guards them. When the larvae hatch, the mother dies and the larvae begin the planktonic cycle of their lives.

Warning: Octopus can bite hard with their beak-like jaws; the poison is not dangerous to humans. However, in the tropical parts of Australia there is a small species of octopus, the blue-ringed octopus. It grows no larger than 12 cm and has conspicuous blue rings. Its poison can kill a human being within two hours.

Polychaetes – Polychaeta

Class Polychaeta, Phylum Annelida. Some 5500 species. Size: up to 20 cm.

Characteristics: Polychaetes occur in widely differing forms. The feature they have in common is a body divided into many segments – the number of segments varies. Bristle worms are rarely to be seen in tropical waters because they usually conceal themselves. Their segmented bodies are fringed at the sides with many sharply pointed bristles or bunches of bristles. When touched, these enter the skin, break off and cause burning pain. The head has several organs of touch, and some species have eyes. Tube worms are fairly common in some tropical waters. Some species build refuges out of material with a lime content, others stick sand and shell fragments together with a rapidly hardening slime or construct a leathery elastic tube from their secretions. The most common species have chosen the most secure refuge – their tubes are to be found in the lime of corals. All species stick one or two rings of tentacles out of their tube or hollow. These provide them with oxygen and food. The tentacles consist of feathery spines, which usually taper as they wind their way upwards in a spiral. They can be found in many colours: red, yellow, blue, white, violet, beige or brown, and often in two colours.

Found: In all tropical to cool seas.

Way of life: In contrast to bristle worms, tube worms are tied to a sessile way of life. Their movements are limited to the short stretch in and out of their tube and the opening out and closing of their tentacles. Without their tentacles they could not survive. They could only have survived in the course of their development if they could pull their tentacles in fast enough. Some species can close the opening of their tube with a small conical stopper which has developed out of one or two tentacles. When feeding the tentacles are held motionless in the current and filter out the plankton, which is guided to the mouth opening by cilia movements.

Food: Plankton and organic drifting matter.

Reproduction: Sexes are separate in almost all species. Sexual and asexual reproduction occurs in many forms. Fertilisation can be by copulation or by the ejection of eggs and sperm into the water. Most larvae develop as plankton.

Christmas tree worm, *Spirobranchus giganteus*; about 2 cm. They like to settle on stone coral and build a tube which later extends inwards. Photo: Thailand

Christmas tree worm, *Spirobranchus giganteus*; each creature has 2 crowns of tentacles which can be pulled in very swiftly when danger threatens.

Fan worm, *Sabellastarte indica*; about 8 cm. Surrounded by sea anemones. Photos: Thailand.

Goose Barnacles – Lepadomorpha

Sub-order Lepadomorpha, Sub-class Cirripeda, Class Crustacea, Phylum Arthropoda. About 800 species. Size: maximum 5 cm.

Characteristics: Only by close observation can one see the slender, tentacle-like legs coming out of the shells and waving rhythmically. The almond-shaped shells have stalks which are firmly fixed to a base.

Found: In all seas, along rocky shores and on movable objects such as driftwood and ships.

Way of life: The movements of the sessile goose barnacles are limited to the opening and closing of their shells and the motion of the legs, which act as filter apparatus for feeding purposes. The legs are equipped with many little hairs along the underside – particles of food adhere to these.

Food: Plankton.

Reproduction: They are predominantly hermaphrodite, fertilising each other in close-packed colonies. The eggs develop within the creatures' bodies: the larvae are planktonic.

Barnacles – Balanomorpha

Sub-order Balanomorpha, Sub-class Cirripeda. Size: up to 3 cm.

Barnacles, *Tetraclita* sp.; about 2 cm. They are common on rocky shores in the tidal zone. Photo: Thailand

Characteristics: Barnacles are sessile crustaceans, hardly recognisable as such. Their bodies are completely surrounded by a mantle which has plates of lime in its outer layer. In danger, or when dry during low tide, the opening is closed by inner shell plates. The rough, often ribbed shells are grey. The legs serve to obtain food, and wave out of the central opening to catch food on fine hairs.

Found: In all seas, mainly on rocky bottom near the surface, but also on turtle carapaces and ship hulls. Rare on coral reefs.

Way of life: The larvae literally 'stand on their heads' attaching themselves to rocks or other solid objects during their transformation into an adult crustacean.

Food: Plankton and organic drifting matter.

Reproduction: As for goose barnacles.

Warning: Take care in surf: all cirripeds have rough, sharp-edged shells.

Shrimps – Natantia

Sub-order Natantia, Order Decapoda, Sub-class Malaco-straca, Class Crustacea, Phylum Arthropoda. Size: maximum length 15 cm.

Characteristics: Most species are good swimmers and live in the open sea. Only a few can be observed in coastal waters. Body build is usually delicate, the tail elongated. The first two or three pairs of the five pairs of legs have pincers.

CLEANER SHRIMP FAMILY: Stenopodidae, Superfamily Penaidea.
A typical feature is the enlarged third pincer-bearing leg pair. Tropical species have conspicuous colouring (white banded with red) and have thick hairs. They have six very long white antennae.

GNATHOPHYLLID FAMILY: Gnathophyllidae, Superfamily Caridea.
Only the first and second pair of legs have pincers; two of the antennae are glassy, transparent, and can hardly be seen, two more are broadened into a leaf shape. Gnathophyllid shrimps are white and have bluish or reddish spots on their shells; the legs are more strongly coloured and banded.

SNAPPING SHRIMP FAMILY: Alpheidae, Superfamily Caridea.
The first pair of pincer-bearing legs is relatively large. They live on the sandy and stony sea bed and are therefore usually inconspicuous in colouring and patterning.

Found: In tropical, calm waters.

Way of life: Cleaner shrimps mainly clean benthic (bottom-dwelling) creatures such as morays, even crawling into their mouths and gills to free them from parasites. Gnathophyllid shrimps live in caves and feed only on starfish, which they locate with their antennae. They crawl along following the trail of the starfish and turn it onto its back in a few seconds. Snapping shrimps live in symbiosis with gobies. They dig a hole into the sandy or stony bottom and allow the goby to share it as a refuge. The continual trickle of sand into the hole forces the snapper shrimp to clean it constantly. In the meantime the goby stays outside on sentry duty. When danger threatens it withdraws into the hole, which the snapping shrimp will only leave "when the coast is clear". The shrimp feeds on small invertebrates (see also gobies, page 182).

Food: See way of life.

Reproduction: Heterosexual: spermatophores are passed on to a freshly moulted female. The eggs are carried under the tail until the larvae hatch. The larvae live as plankton for 1-3 months, and their development goes through several stages.

Banded cleaner shrimp, *Stenospus hispidus*; about 8 cm without antennae. They usually clean benthic (bottom-dwelling) fish. Photo: Maldives.

Gnathophyllid shrimp, *Hymenocera elegans*; about 5 cm. They feed exclusively on starfish, which they can quickly turn onto their backs. Photo: Maldives.

Snapping shrimp, *Alpheus* sp., with goby; about 6 cm. They live in symbiosis with the goby *Amblyeleotris steinitzi* in a common hole. Photo: Thailand.

Spiny lobsters – Palinuridea

Superfamily Palinuridea, Order Decapoda, Sub-class Malacostraca, Class Crustacea, Phylum Arthropoda. More than 8000 species. Size: about 45 cm.

Characteristics: They can easily be distinguished from other families of crustaceans by their extremely long antennae-like feelers which extend far beyond the body and by the lack of large pincers. At the end of the long tail is a caudal fin which they use for swimming. The tail is, however, usually folded under; it is divided into several segments. The body of the creature is covered by an external skeleton known as its armour or (in its frontal regions) carapace. The sharp protrusions and spines found on large parts of the body are the spiny lobster's only defence. They have three pairs of long running legs, which are vertically striped.

Found: In all warm and cool seas, usually in rocky areas with many caves.

Way of life: Spiny lobsters are found almost exclusively on firm ground: this can be the roof of a cave on which they hang. They are skilled and fast runners and can easily escape their enemies. Often, however, they take a "wait and see" attitude and try to use their long feelers to find out about their opponent. This is frequently fatal. Since diving has become a popular sport their numbers have declined considerably. Even fishermen today are buying diving equipment and further decimating the already scanty population. The natural enemy of the spiny lobster is the octopus, which kills them with its powerful beak, biting through their armour. As with all crustaceans, the outer skeleton is hard and consists of chitin and lime deposits. It does not grow with the animal and so crustaceans have to renew their shells several times in the course of their lives. The old shell, now too small, is then left together with its parasites and growths. The new outer layer underneath is still elastic and lies in folds, as it is larger than the old. During this phase the creature is unprotected and lives in hiding. All crustaceans grow only during moulting.

Food: Gastropods, mussels and carrion, also algae.

Reproduction: These heterosexual creatures reproduce by the male giving a freshly moulted female a sperm packet (spermatophore). The female carries the eggs under the tail, until the larvae begin their life as plankton. Their complex development goes through several stages and can take up to three months.

Spiny lobster, crayfish, *Palinurus* sp.; about 40 cm without antennae. They have unfortunately become very rare. Photo: Maldives.

Spiny lobster, *Palinurus versicolor*; about 30 cm. Often to be seen on the roofs of caves. Photo: Thailand.

Hermit Crabs – Paguridea

Superfamily Paguridea, Order Decapoda. Up to 18 cm long (without shell).

Characteristics: They live exclusively in the shells of gastropods, which they search out according to their size. The shells not only protect their soft and vulnerable rear, but they can also withdraw right into it when in danger and close the opening with their largest pincer. The pincer-bearing pair of legs is usually asymmetrical, as is the abdomen of the creature, which may coil to the left or the right according to the shape of the gastropod shell. Hermit crabs angled to the left, therefore, can never live in a shell that turns to the right. The crab holds itself in the shell with its final pair of legs, which have a rough, non-slip surface. It only leaves the shell to look for a larger one. The long-stalked eyes are prominent, surveying their surroun-dings after emerging slowly. The shells are sometimes overgrown with sponges or settled by parasite anemones.

Found: Widespread in all seas, in water that is not too deep. The hermit crabs which live on land are descended from the "sea hermits" but have atrophied gills and would drown in water.

Way of life: They are mainly nocturnal, but can be seen by day in coral reefs. They are excellent climbers, and it is not unusual for them to clamber about on stone coral. It can happen that they get so thoroughly stuck in the confusion of coral branches that they cannot get free. Even under the circumstances the crab will not leave the shell, as it would soon, with its worm-like abdomen, fall prey to fish. Hermit crabs sometimes live in symbiosis with parasite anemones which protect the crab against enemies. "Parasite" is an unfair name for these anemones, as the crab actively persuades them to settle on its shell. In some species the hermit crab grabs the anemone quite roughly with its pincers and presses it onto its shell home until it is firmly attached. In other cases the crab has been observed curling its legs about the anemones and trying to persuade it by stroking to slowly remove its foot from its base and change over to the crab's shell. If a crab changes shell, it takes its anemone with it.

Food: Omnivore, can be predatory.

Reproduction: The larvae undergo two stages as plankton before they settle.

Diogenes hermit crab, *Dardanus diogenes*; about 7 cm. Small parasite anemones are growing on the shell. Photo: Thailand.

Small hermit crab, *Paguristes* sp.; about 6 cm. In the shell of a cowry. Photo: Thailand.

Crabs – Brachyura

Sub-order Brachyura, Order Decapoda. Size: generally up to 30 cm maximum. About 4500 species.

Characteristics: All species have a compressed body which is often broader than it is long. The abdomen is atrophied and always folded forward under the thoracic carapace: in females it is broader and serves to protect the young. Crabs are divided into a number of families such as Porcellanidae (porcelain crabs), Hippidae (mole crabs), Dromidae (sponge crabs), Calappidae (box crabs), Raninidae (spanner crabs), Leucosidae (pebble crabs), Majidae (spider crabs), Portunidae (swimming crabs), Xanthinidae (dark-fingered coral crabs), Ocypodidae (ghost and fiddler crabs), Grapsidae (shore crabs), and several others. The first of the five pairs of legs is always equipped with pincers. These can be symmetrical or asymmetrical. The length of the legs varies greatly. Swimming crabs have a broader pair at the rear, which they use as "oars" to swim with. The outer skeleton is often equipped with spines and sharp edges, other crabs are thickly "furred". Colouring and patterning vary greatly.

Found: In all seas, mainly in coastal regions, also on land.

Way of life: While various species of crab can be observed in the water and on land along rocky shores on beaches, those species which live exclusively in the water lead a hidden life. Swimming crabs mainly live in the open sea. Crabs are conspicuous because of their gait, which is unusual in the animal kingdom. They walk sideways, although they can also walk forwards and backwards. They are more skilled at climbing than crustaceans with a long tail, which may be useful for swimming but is a hindrance when walking. The crab's pincers are used for grasping food, but are also powerful defensive weapons. They use them to scrape algae off rocks or to break up large pieces of food. Spider crabs have an unusual method of camouflage; they use their pincers to place algae or sponges on their carapaces, and prefer to stay in spots where they are well camouflaged. If they are moved to a place where they are not well camouflaged they will very soon start to settle suitable creatures and algae onto their shells.

Food: Omnivorous. Cannibalism is frequent.

Reproduction: Mainly heterosexual; the female carries the eggs under her abdomen until they hatch. Males can be recognised by their slimmer abdomens. The larva's plankton life can last for up to 3 months and goes through three different stages.

Swimming crab, Family Portunidae; about 10 cm. The oar-like back legs are raised. Photo: Thailand.

Rock-dwelling crab, *Atergatis subdentatus*; about 15 cm. During the day they remain hidden in caves of clefts. Photo: Thailand.

Echinoderms: Echinodermata

All echinoderms have a system of tubes filled with sea water running through their bodies which can change the size of elastic hollow spaces by changes in pressure. This water-vascular system provides the locomotion of sea cucumbers, sea urchins and sea stars; in crinoids and brittle stars it is used to transport food.

Crinoids – Crinoidea

Class Crinoidea, Sub-phylum Pelmatozoa, Phylum Echinodermata. Some 620 species. Size: up to 30 cm in diameter with arms outstretched.

Characteristics: They have a tiny, cup-shaped body, generally with 10 or 20 feathery arms which act as filters to obtain food. Their numerous slender legs, known as cirri, grasp a base. The number of cirri varies from 18 to 20. They are composed of many jointed, linked limbs. When feeding the arms are stretched out in a rayed pattern and are curled up when inactive. Crinoids are often splendidly coloured. There are single-coloured and patterned species – vertical and horizontal bands predominate.

Found: In tropical and temperate seas, preferably on current-exposed, raised sites, such as on horny corals or sponges.

Way of life: Crinoids can climb well using their slender cirri. However, they can also swim, moving their feathery arms in an elegant rowing motion. They are filter feeders, catching plankton by facing their arms into the prevalent current. When caught, the plankton is transported to the mouth in the water-vascular channels on the upper side of the arms. Mouth and anus are on the upper side of the body. If the chosen site is favourable and they are not disturbed, they may often stay for weeks in the same spot. They are active by day and night. The skeleton (test) of the creatures consists of movable linked plates of lime, which are very brittle. The arms break off easily, but can regenerate.

Food: Plankton.

Reproduction: The sexes are separate but have no external sexual characteristics. Fertilisation is by the release of sperm into the water. The eggs stay attached to the female's arms until the larvae hatch. The larvae, after a short stage as plankton, sink to the sea bed and anchor themselves with a stalk until they are fully developed.

Crinoids, Family Comasteridae (feather stars); about 25 cm. They can be found in many different colours. Photo: Thailand.

Feather star, *Oxymetra erinacea*; about 20 cm. They use their cirri to climb to raised sites, but can also swim, using their feathery arms. Photo: Thailand.

White feather stars on gorgonian fan coral; about 40 cm. Some species have relatively few arms. Photo: Philippines.

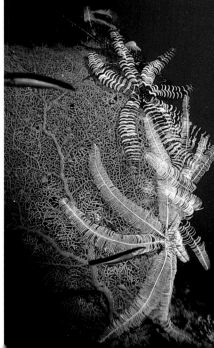

Sea Cucumbers – Holothurioidea

Class Holothurioidea, Sub-phylum Eleutherozoa, Phylum Echinodermata. Size: usually up to 50 cm. Giant sea cucumbers can reach up to 3m.

Characteristics: They have a longish, cucumber-shaped body, which gives them their name. In contrast to other echinoderms they have a soft body as the lime skeleton contained in the skin has atrophied into microscopically small parts. On the underside there are three rows of tube feet with which they can attach themselves by suction. The remaining two or five rows typical of this phylum can be found in some species in the form of retractable warts along the back or have atrophied altogether. The mouth and anus lie each at one end of the body. In some species the tube feet have developed into retractable tentacles which can be very branched and which are used in feeding. Sea cucumbers like to hide their bodies in cracks, with only the wreath of tentacles showing out of the ground. They can then be mistaken for corals. The leathery skin of the creatures varies in structure: in some species it is relatively smooth, in others it is warty, wrinkled or studded with bumps, usually inconspicuous in colour. In some areas there are magnificently coloured species, which may also be patterned.

Found: In tropical to cooler seas.

Way of life: Sea cucumbers live on the sea bed; they have no chewing mechanism; the tentacles of some species can be used as an aid to scraping, catching and placing food into the mouth. Sea cucumbers ingest large amounts of sand with their food; the particles of food contained in the sand are digested. The animals move slowly by using their tube feet or by contracting and stretching their bodies. Some species protect themselves against predators by burying themselves in soft ground by day and only coming out to search for food at night. When attacked, some species can extrude white sticky threads, the Cuvier threads, which can entangle predators.

Food: Algae, detritus or plankton.

Reproduction: Sexes are usually separate but cannot be differentiated externally. There are also hermaphrodites. Eggs and sperm are released into the water. Some species can assume a relatively upright posture when doing so. Development of the larvae generally goes through two stages and only lasts 9-14 days.

Sea cucumber, *Bohadschia graeffei*; about 35 cm. Feeds mainly on detritus. Photo: Maldives.

Giant sea cucumber, *Polyplectana kefersteini*; up to 3 m. It catches food in its feathery tentacles and guides it into its mouth. Photo: Maldives.

Sea Urchins – Echinoidea

Class Echinoidea, Sub-phylum Eleutherozoa, Phylum Echinodermata. Some 860 species.

Characteristics: Sea urchins are divided into two sub-classes. The regular sea urchins are symmetrical in construction and the irregular are asymmetrical. The latter are not described here. Regular sea urchins have almost spherical bodies and are slightly flattened underneath. They are equipped with movable ray-like spines which are usually long on the back and shorter underside. The mouth is in the centre of the underside and has five relatively large concentrically placed teeth. The anus is on the upperside and in some species it ends in a visible blister. In this group the plates of lime within the skin form a carapace. They are divided into five bands running from the mouth to the anus between which are found the tube feet and through which water circulates. Only a few species have thick, blunt spines. The colours vary greatly; dark, inconspicuous colours predominate, some species are intensely red, blue or pale.

Found: In all seas, from shallow water to great depths, on all types of sea bed.

Way of life: Sea urchins live on the sea bed and move forward slowly using their sucker-studded tube feet or their spines. They are nocturnal and mostly hide during the day. When they occur in greater numbers and there are not enough hiding places, they group together in open spaces. Enemies which have been sensed by light-sensitive organs can be chased away by active movements of the spines. Some small specialised species can dig into rock, if it is not too hard, with their teeth and then live a sessile life in these holes, as long as they can obtain enough food from particles drifting in on the current. Sea urchins generally live as grazers, scraping growth from the sea bed. Their enemies are crabs, sea stars and larger fish, especially triggerfish.

Food: Algae, micro-organisms, organic material and sinking matter.

Reproduction: The sexes are always separate. Eggs and sperm are ejected into the water. The plankton stage of the larvae lasts for 4-6 weeks.

Warning: Sea urchin spines break off easily and remain in a wound. Some sea urchins are poisonous and have little blisters of poison on their spines – they resemble berries. When a spine penetrates the skin, the poison is introduced.

Sea urchin, *Mespila globulus*; about 7 cm. This species often uses foreign bodies to disguise itself. At the bottom right of the photograph some of the little tube feet can be seen. Photo: Thailand.

Slate pencil urchin, *Heterocentrotus mamillatus*; about 20 cm. These nocturnal animals wedge themselves into holes, using their thick spines, by day. Photo: Red Sea.

Long-spined sea urchin, *Diadema setosum*; about 25 cm. If there are not enough hiding places, they collect in groups as a protection against enemies. Photo: Red Sea.

Ophiurids – Ophiuroidea

Class Ophiuroidea, Sub-phylum Eleutherozoa, Phylum Echinodermata. About 1800 species.

ORDER EURYALAE: (Basket Stars). Size: up to 1.5m in diameter.

Characteristics: Basket stars have a small, compressed body with five main arms which branch extensively; each one can have several thousand tendril-like tips with little hooks to catch floating particles. The creatures have a mouth but no anus. Colour varies.

Found: In tropical and temperate sees.

Way of life: They are exclusively nocturnal animals, and if caught in light they roll their arms up. They spend all day in this position in a hiding place. When darkness falls they come out and crawl, using their arms, to their customary raised site. There they spread their arms broadside on to the current, forming a huge fan. The plankton filtered out from the current is guided to the mouth via the water-vascular system.

Food: Exclusively plankton.

ORDER OPHIURAE: (Brittle Stars)

Small, compressed body disc with unbranched arms edged with many spines. The mouth is situated on the underside of the body disc, there is no anus. The water-vascular channels are covered with skeletal plates. Brittle stars can be found in many colours; patterning also occurs frequently.

Found: In tropical and cooler seas; under stones, in cracks, on sponges and coral.

Way of life: Brittle stars are not often seen in the tropics. The filter feeders can occasionally be seen at favourable points in the current, usually on coral, while the grazers look for protection from predators, e.g. starfish, in the narrowest cracks. They can move their arms quickly with a snake-like motion and push themselves forward fairly rapidly in this manner. Missing limbs can be regenerated.

Food: Plankton, sinking matter and small creatures, e.g. worms, crustaceans, see urchins and other soft-bodied animals.

Reproduction: The sexes are generally separate without external sexual characteristics. Hermaphrodites are rare. Some brittle stars also bear live young. Planktonic larval stage usually lasts 5-9 weeks. Also asexual reproduction by division.

Basket star,
Astroboa nuda; up to 1.5 m. When darkness falls they climb to a raised position and catch plankton. Photo: Red Sea.

Brittle star,
Ophiothrix sp.; about 15 cm. Some species live on sponges or coral, while others hide under stones and in cracks. Photo: Thailand.

Ascidians – Ascidiacea

Class Ascidiacea, Sub-phylum Tunicata, Phylum Chordata. Some 2000 species. Size: up to 30 cm, but usually only around 5 cm.

Class Thaliacea also belongs to the tunicates.

Characteristics: They look more like a sponge than the closest relative of the vertebrates. The body is enclosed in a jelly-like mantle, the tunica, which varies in shape. There are solitary forms which look like a pot-bellied bottle, whereas other species live in colonies; these are mainly very small. The mantle has two openings; one for water coming inwards and one outwards. The inwards opening is always at the end and is often extended in a pipe-like manner. These openings, important for feeding and breathing, can be closed if the creature is disturbed. A few species have a transparent mantle, so that some of the animal's inner organs are visible. Most of the space inside is taken up by a basket-like structure, the branchial sac or pharynx. Many species have an outer surface which resembles an inorganic substance and are not easily recognised as living matter. These thick-walled species are often colonised by other creatures. Many ascidians are magnificently colourful. Patterns are rare.

Found: Ascidians almost all live sessile lives, but some are swimmers. They live in all seas and can be found at every depth.

Way of life: Despite their sessile character, ascidians do not depend on the current, but create a flow of water through their own bodies. This means that they belong to the current feeders. The current through their bodies is created by the movement of the cilia in the branchial sac. The branchial sac serves two purposes – the absorption of food and of oxygen. It is surrounded by a space, the atrial cavity. Among other sensor organs, the ascidians have touch receptors which cause the body openings to contract if the creature is touched.

Food: Organic particles, plankton and bacteria.

Reproduction: Colony-forming species mainly by budding; solitary species are hermaphrodites which eject eggs into the water via their outward openings. There are also viviparous species which are fertilised internally by sperm brought in on the water flow. Self-fertilisation does not occur. The fully developed larvae are tadpole-like in form and can swim. The larval stage is very short; the creatures do not eat during this stage. The chorda, which makes them the vertebrates' closest relatives, is only present in the larvae.

Common solitary ascidian, *Polycarpa* sp.; about 5 cm. They create a current which lets water flow in at the end opening and filter out the plankton. Photo: Philippines.

Colonial ascidian, *Clavelina coerulea*; individual animals about 2 cm. These sessile creatures can move forwards very slowly. Photo: Thailand.

Nurse and Carpet sharks – Orectolobiformes

Order Orectolobiformes, Super-order Selachoidei, Class Chondrichthyes. 33 species. Size: from about 1 m to over 12 m. The whale shark is the largest of all fishes.

Characteristics: Generally a heavy body and blunt, slightly flattened head with two barbels close to the nostrils. The two dorsal fins, clearly separated, are set relatively far back. In most families they are of about the same size (except for the zebra and whale sharks). All species, except for the whale shark, have a long, flat tail.

Found: Whale shark – in the open sea of tropical oceans, occasionally appears near the coast or along reefs. All other species - shallow coastal waters in the tropics; wobbegongs are only found around Australia and Japan.

Way of life: Whale sharks move very slowly when swimming but still achieve a considerable speed. When they meet divers they often react with curiosity and come to within 1-2 m. They sometimes spend long periods of time near divers, as long as they are not touched. Whale sharks are probably active by day and by night. All other families are nocturnal and rest by day; zebra sharks on the sandy bottom, nurse sharks in caves. They almost always sleep in the same cave, but are very sensitive to disturbance. A gentle touch is all it takes for them never to return to their customary place, although they show almost no reaction to being disturbed at the time. At night they search the sand for food, which they grind up with their blunt teeth.

Food: Whale sharks filter plankton; all others eat fish and invertebrates.

Reproduction: Internal fertilisation takes place, with the male inserting a penis-like clasper into the cloacal cavity of the female. Sperm flows through a channel within the clasper. Nurse sharks are oviparous, they lay their eggs on the sea bed. (The process is not fully recorded for whale sharks). The young develop within an elastic egg membrane and feed from a large yolk sac. After about 9 months (the time span varies according to water temperature) the young hatch; as soon as the yolk sac is used up, they begin to search for food.

Warning: When provoked, nurse sharks can bite and hold on like bulldogs. It is almost impossible to get out of the water with a 3 m long nurse shark attached.

Nurse shark, *Nebrius concolor*, up to 3.2 m. Is found in caves by day; reacts very definitely to being touched and will never come back to its customary place. From the Red Sea to Tahiti. Photo: Thailand.

Zebra shark, *Stegostoma varium*; up to 2.3 m. Likes to rest by day on sandy sea bed near reefs. Red Sea to Samoa. Photo: Thailand.

Whale shark, *Rhincodon typus*; up to 12.8 m, which makes it the largest of all fishes. Harmless plankton eater. All tropical seas. Photo: Thailand.

Sharks – Carcharhinidae

Family Carcharhinidae, Order Carcharhiniformes. 48 species. Size: 70 cm to about 3 m.

Characteristics: This family has the typical shark form: spindle-shaped body, head relatively pointed, five gill openings, first dorsal fin large, second small. Single-coloured, paler on underside.

SPECIES LIVING NEAR REEFS: Grey reef shark, broad black band bordering the tail fin, white border on the upper half of the first dorsal fin. Whitetip reef shark, greyish-brown, with bright white tips to the first dorsal fin and tail fin. Blacktip shark, pale beige, with noticeable black tip to the first dorsal fin and the lower part of the tail fin.

Found: In tropical seas world-wide.

Way of life: The so-called reef sharks calmly patrol their territories along the reefs. Only in certain circumstances will they demonstrate how fast they can swim; when hunting, when they are alarmed or when apparently attacking. Although they can be dangerous to humans, they mostly avoid them and keep well away. In some areas where they are common they will leave their territories if diving takes place there regularly. If you dive in a completely untouched area, you will have to count on many sharks appearing suddenly; they remain very curious, until they have become used to divers. Whitetip reef sharks often rest by day on sandy bottoms. They are most commonly found at depths of 10-25 m. Juvenile blacktip sharks can be seen in very shallow water. They occur in large numbers in the lagoons in some areas, and occasionally they will swim very close to the shore at depths of only 20 cm, with their dorsal fins appearing above the water. Adult specimens, however, are very rarely met. Grey reef sharks prefer atoll channels with strong currents and depths of 25-50 m.

Food: Fish, squid, crustaceans.

Reproduction: Grey reef sharks leave their territories for a certain time every year. When they return the females have many bite wounds from mating. The females are encouraged to mate by these "lovebites". Copulation takes place "belly to belly". The males of some species bite the pectoral fins of the females in order to hold on. Often all the fins are torn. Carcharhinidae are viviparous, i.e. they bear live young. During and after birth the females have an inhibition against biting, so that they do not eat their young. At this time the males are to be found in other areas.

Grey reef shark, *Carcharhinus amblyrhynchos*; up to 2.3 m. Fairly common in the Maldives in certain places. Photo: Maldives.

Whitetip reef shark, *Triaenodon obesus*; up to 1.7 m. A widespread species which is almost always found as single individuals. Red Sea and the entire Indo-Pacific. Photo: Maldives.

Rays – Batioidei

Superorder Batioidei, Class Chondrichthyes. 470 species. 40 cm to 6.7 m across.

Characteristics: Body extremely compressed vertically, gill, mouth and nostril openings usually on the underside. The form varies greatly.

Found: World-wide in tropical and sub-tropical seas; mantas and eagle rays usually swim in open water, all other species live close to the sea bed.

Way of life: Rays have very different swimming techniques: mantas and eagle rays are fast swimmers, moving their much enlarged pectoral fins up and down in a wing-like fashion. Electric rays, sawfishes, and guitar fishes swim using their tail fin. True rays and stingrays move the fin fringing their bodies in a wave-like manner. Rays have an opening just behind each eye through which they suck in water and guide it through the gills. Water re-emerges through the gill openings on the underside. This ability makes it possible for them to camouflage themselves by burying themselves in sand. They often lie in very shallow water. If they are overlooked by people wading, this can have drastic consequences. They rely on their good camouflage and rarely take flight. Rays are usually found as single individuals; it is only eagle rays who can be seen in small groups and mantas who appear in large schools, sometimes of 40 or more individuals.

Food: Mantas: plankton and shrimps; all the other species eat fish and invertebrates living on the sea bed.

Reproduction: As with sharks, fertilisation in rays takes place internally, with the male inserting a clasper into the female. Apart from the true rays, all members of this order bear live young.

Warning: Stingrays have one or two long, poisonous stings with many barbs which can cause very painful wounds. The stings are on top of the tail behind the body disc. When the creatures feel threatened they raise their tails and hold them in a threatening position above their bodies. If you approach them further, you run the risk of being struck by the whip-like tail or, even worse, by the poisonous stings. The sting often breaks off in the wound and can then only be removed surgically, due to the many barbs. Eagle rays can also sting. Electric rays have electric organs which can produce shocks of over 200 volts.

Manta ray, *Manta birostris*; up to 6 m with fins spread. Largest species of ray, harmless plankton feeder. All tropical seas. Photo: Djibouti.

Spotted eagle ray, *Aetobatis narinari*; up to 2.3 m with fins extended. Mainly in open water with currents. All tropical seas. Photo: Maldives.

Black-spotted stingray, *Taeniura melanospilos*; up to 1.65 m. They live exclusively on the sea bed. Photo:: Maldives.

Morays – Muraenidae

Family Muraenidae, Order Anguilliformes. About 100 species. Size: from 20 cm to almost 4 m in length.

Characteristics: Extremely elongated, scaleless body, which is very muscular and strong. They have neither pectoral nor pelvic fins. The dorsal, tail and anal fins have fused to form a fringing fin. If only the creature's head is visible, it can be identified by the fact that it has no gill covers. There is a clearly visible gill opening behind the head on either side. The mouth is very large and is continually being opened and shut. This movement is intended to help oxygen absorption and pumps water through the gills. The eyes are set very far forward, well in front of the corners of the mouth. The nostrils are equipped with small protruding tubes.

Morays have a very good sense of smell which allows them during their nocturnal hunts to detect such elusive prey as sleeping, uninjured fish. However, their sight is poor.

The creatures can probably hear very well. Most species are inconspicuous in colour and thereby well camouflaged; only a few are distinctively patterned or coloured. Many have a fearsome set of teeth. The big, dagger-like fangs for catching prey are in the centre of the upper jaw. Other species have many smaller teeth or pebble-like grinding teeth which are not visible.

Found: World-wide in warm and temperate seas; in caves, crevices and holes which they rarely leave by day.

Way of life: Morays are largely active by night or at dusk. They withdraw by day. They are only to be found near the sea bed, live in their own territory and have one or probably more favourite spots. These can be as close as 200 m and are visited at irregular intervals. Morays can only occasionally be seen swimming freely, when they are on the move from one coral block to another, close to the sea bed. However, this behaviour has changed in some areas, as morays have been fed by divers and, once they have become accustomed to such feeding, swim up to divers.

Morays, with their snake-like swimming movements, are not among the fastest swimmers. They are usually found as individuals, but do gather in groups when young. Sometimes two different species will live in one hole. Frequently only the head is visible. Their breathing movements are often interpreted as threatening. When a moray makes regular breathing movements, this is a sign that it does not feel nervous.

Giant moray, *Gymnothorax javanicus*; up to 2.4 m. The best-known and most common species. Not aggressive, as long as they are not fed by divers. Red Sea to Tahiti. Photo: Maldives.

False honeycomb moray, *Gymnothorax favigineus*; up to 2.5 m. The white lines on the true honeycomb moray are narrower. Indian Ocean to West Pacific. Photo: Thailand.

A moray which feels threatened or is startled will draw back slightly and open its mouth wide. It will stay in this position until it realises that the danger is past. The widely held belief that morays are aggressive only applies to a few species such as the tiger moray *Uropterygius tigrinus* and the viper moray *Enchelynassa canina*. As a rule morays will only bite if they are caught, wounded or cornered. One further exception is the genus *Echidna*, which hunts in very shallow water and even approaches the shore. There are reports of attacks on humans in these circumstances. It probably happens because the animals become desperate if their route back into deep water is blocked. If the same species are met with in deep water they are quite peaceable.

Food: Fish, squid, crustaceans; dead or alive.

Reproduction: Moray spawn floats on the surface. The larval stage is very long; the larva is a mere 1 cm long after one week.

Warning: Morays which have been fed for a long period of time no longer hunt for themselves and will therefore approach divers. They will snap at anything that smells or looks like food. Accidental bites are not poisonous, as used to be believed, but can cause infections which take a long time to heal. Only the blood of morays is poisonous if it gets into the human bloodstream. Full grown morays should never be eaten, as this may lead to ciguatera poisoning.

White-eyed moray, *Siderea thyrsoidea*; up to 65 cm. Sometimes found in groups of up to 50 individuals. Thailand to Tahiti. Photo: Thailand.

Yellow-headed moray, *Gymnothorax fimbriatus*; up to 80 cm. Photographed inside a beaker sponge. Red Sea to Hawaii and Tahiti. Photo: Thailand.

Ribbon eel, *Rhinomuraena quaesita*; up to 1.2 m. Juveniles are black and yellow up to a length of 65 cm, males from 65-90 cm are blue and yellow and females up to 1.2 m can be yellow. East Africa to Tahiti. Photo: Maldives.

Geometric moray, *Siderea grisea*; up to 70 cm. A species found only in the Red Sea. Photo: Egypt.

Catfishes – Plotosidae

Family Plotosidae, Order Siluriformes. Of the c. 2000 species only a few are marine. Size: Maximum 30 cm in length.

Characteristics: Slender body, slightly compressed head. Four pairs of barbels around the mouth. The first dorsal fin is very short and high. The second dorsal fin is fused with the tail and anal fins to form a bordering fin. The scaleless body is dark. The young have horizontal stripes.

Found: From the Red Sea to the Pacific.

Way of life: Adults live as individuals and stay well hidden. The young keep close together to form a "rolling ball" over sandy bottoms or seaweed.

Food: Fish, crustaceans, molluscs and sinking material.

Reproduction: Only in the case of *Plotosus lineatus* have the males been observed building nests under stones and guarding the eggs.

> Warning: These catfishes have a toothed poisonous spine on the pectoral and first dorsal fins. Injuries can be very painful and the pain may last for over two days. The sting of a young catfish no more than 3 cm long is supposed to hurt like a wasp sting. However, catfish never attack of their own accord.

Lizardfishes – Synodontidae

Family Synodontidae, Order Aulopiformes. Some 50 species. Size: up to 35 cm long.

Characteristics: Cylindrical body which tapers slightly towards the back; the mouth is very large, with many small teeth. They have only one dorsal fin. All fins are at least partly transparent and are therefore not easy to see. Most species have easily recognisable scales. All lizardfishes are well camouflaged. The background shade is usually a sand colour; the darker patches have no clear symmetrical forms. The patterning is important for the identification of the species.

Found: In shallow tropical and temperate seas.

Way of life: They lie motionless on the bottom, resting on their pectoral fins, in wait for prey. If they are startled they swim for a short way and then lie back down motionless on the sea bed.

Food: Small fish.

Reproduction: Little is known.

Striped eel catfish, *Plotosus lineatus*; up to 30 cm. Fully grown adults are a uniform brown in colour. The young live in groups. Red Sea to Samoa. Photo: Thailand.

Graceful lizardfish, *Saurida gracilis*; up to 30 cm. They lie in wait for prey, motionless, on sandy or stony sea beds. Red Sea to Tahiti. Photo: Egypt.

Trumpetfishes – Aulostomidae

Family Aulostomidae, Order Syngnathiformes. Only four species known. Size: maximum 75 cm.

Characteristics: Slender, cylindrical, extremely elongated body. Under their scales is a network of small bones which form a protective armour. The head has a long, forward-pointing snout with a mouth at the end. The second dorsal fin and anal fin are well developed. All trumpetfishes are well camouflaged by colour and pattern, with one exception: *Aulostomus chinensis* also occurs in a conspicuous shade of yellow (see photo above).

Found: In tropical regions in the Atlantic, Indian and Pacific Oceans, usually among corals.

Way of life: Trumpetfishes are steady swimmers and live singly or in pairs. They are often seen in hiding upright or at an angle amid coral. When hunting they use cunning. They swim close above a peaceful fish so as not to be easily seen by potential victims. When they are close enough they shoot forwards and snatch the surprised victim. Yellow trumpetfish prefer to "ride" above yellow rabbitfish.

Food: Small fish and crustaceans.

Cornetfishes – Fistularidae

Family Fistularidae, Order Syngnathiformes. Only a few species. Size: about 1.5 m in length.

Characteristics: Extremely slender long body, tapering towards the back. The head is also extremely long and is only broadened by the eyes. All the fins except for the pectoral fins are set far back and are symmetrically opposed; first dorsal fin with the pelvic fin and second dorsal fin with the anal fin. The tail fin is a thin thread-like addition which elongates the body further. The scaleless skin is bluish-green and camouflages the animal well in open water. The fishes can camouflage themselves quickly with a pattern of dark, greyish-brown blotches.

Found: In tropical regions of the Atlantic, Indian and Pacific Oceans.

Way of live: Cornetfishes live alone and also in small groups. They hunt just below the surface. Sometimes they allow themselves to drift like flotsam up to shoals of small fish, until they are close enough to their victims.

Food: Fish and crustaceans.

Trumpetfish, *Aulostomus chinensis*; up to 65 cm. Yellow colour variation (see also centre photo). East Africa to Hawaii. Photo: Maldives.

Trumpetfish, *Aulostomus chinensis*; up to 65 cm. Colour varies from yellow (see photo above) to green or brown. Typical characteristics are the white spots at the end of the body. East Africa to Hawaii. Photo: Thailand.

Cornetfish, *Fistularia commersonii*; up to 1.7m in body length. The thread-like tail is conspicuous. Red Sea to Hawaii. Photo: Maldives.

Seahorses and Pipefishes – Hippocampae and Syngnathinae

Sub-families Hippocampae and Syngnathinae, Family Syngnathidae, Order Syngnathiformes. About 200 species. Size: Seahorses up to 30 cm, pipefishes up to 40 cm in length, but most are considerably smaller.

Characteristics: Characteristics this group have in common are: tubelike snout with small mouth at the end; no gill covers; body covered with numerous hard bony rings, angular in structure. Colour and patterning vary greatly.

Seahorses always have a horse-like head bent forwards onto the breast. They have dorsal and pectoral fins; the anal fin is tiny or missing. The "tail" has become a long prehensile limb.

Pipefishes are slender elongated fish; their heads always lie more or less in a line with their bodies. Most species have a round tail fin; only a few have a prehensile tail like the seahorses.

Found: Seahorses are very rare in coral reefs due to the strong currents.

Way of life: Seahorses are not only the most unusual of bony fishes in appearance; their upright method of swimming is just as intriguing. They are slow swimmers and can hold on to plants or similar suitable growth with their prehensile tails. Their body movements are limited by their armour; only the tail is very mobile. Food is brought to the mouth by the current.

Pipefishes can swim better due to their round tail fin and can be observed in many tropical regions. They prefer protected sites such as overhangs, caves and bays in the reef.

Food: Zooplankton and benthic invertebrates.

Reproduction: Both sub-families bear live young in an unusual manner: when mating, the female does not receive sperm from the male, but lays eggs either into the breeding pocket (seahorses) or in a fold on the belly (pipefishes) of the male. With some pipefishes the eggs are only stuck on to the belly. The young hatch on or in the male. The breeding pocket is cleaned and prepared for the next batch of eggs. A little later the female lays more eggs. This reproductive cycle usually takes place three times a year.

Yellow seahorse, *Hippocampus kuda*; up to 25 cm. Very rare in coral reefs, prefers seaweed. Red Sea to Hawaii. Photo: aquarium.

Banded pipefish, *Doryramphus dactyliophorus*; up to 18 cm. Red Sea to Tahiti. Photo: aquarium.

Flatheads – Platycephalidae

Family Platycephalidae, Order Scorpaeniformes. About 60 species. Size: up to 80 cm; there is only one species, in Australia, which can grow to 1.2 m.

Characteristics: Elongated, slender body with flattened head similar in shape to a crocodile's. The name "mail-cheeked" comes from a peculiarity of the order's anatomy - a bony plate below the "cheek" between the orbital and the gill cover. The large, round button-like eyes are covered with a net-like organ which can contract and stretch. This protection against light comes into use when bright light enters the water; these are probably the only fish to possess a feature resembling an eyelid. They are predators and as such have very large mouths with protruding lower jaws. The two dorsal fins are clearly separated; pectoral and pelvic fins are well developed and partly overlap. The tail fin, when at rest, is folded flat. Colour and irregular blotchy patterns are well adapted to the sea bed. There are also many backward-pointing spines and bony plates on the head which, together with the "net" over the eyes, camouflage the creature perfectly. The body is covered with small scales.

Found: Exclusively in the tropical waters of coastal areas, from the Red Sea to the Pacific; on sandy, muddy or stony beds, most frequent in shallow water up to a depth of 20 m. Relatively rare in the Indian Ocean.

Way of life: These predators that ambush their prey cannot swim for long periods. They prefer to frequent sandy and stony sea beds. They are not so well camouflaged in sandy areas, therefore they will dig themselves in completely so that only the eyes and mouth can be seen. They rely on their camouflage and lie still for a long time when approached. At the last moment they will flee a few yards and instantly dig themselves back into the sand. They lie in wait for prey until a victim approaches close enough. Fish which swim past at a distance of 1 m or less are seized with a "leap" at lightning speed. Flatheads are mostly observed singly, occasionally in pairs. Crocodile fish sometimes live in brackish water.

Food: Fish, crustaceans, worms and molluscs.

Reproduction: Largely unknown.

Spotted flathead, *Papilloculiceps longiceps*; up to 1 m. Relatively common in the Red Sea; they like to bury themselves in the sand. Red Sea to Japan. Photo: Egypt.

Longsnout flathead, *Platycephalus chiltonae*; up to 22 cm. Red Sea to Tahiti. Photo: Thailand.

Scorpionfishes – Scorpaenidae

Family Scorpaenidae, Order Scorpaeniformes. About 300 species. Size: 4-50 cm according to species.

Characteristics: Powerful perch-like body, large head; many short, generally backwards-pointing spines. Eyes and mouth are relatively large. The dorsal fin is continuous, the pectoral fins large; the whole body is covered with scales. The colour varies greatly: most species have a basic reddish colouring, which can however be adapted to the surroundings. Irregular blotchy markings and countless attachments to the skin cover the body and perfect the camouflage of this predator that ambushes its prey. Some also have tree-like growths around the lower jaw, in order to camouflage the head even better. A widespread species, *Scorpaenopsis diabolus* needs special mention. Its English name is well-earned – false stonefish. It is grey, sometimes greenish and occasionally reddish in colour and is often taken for a stonefish because of its resemblance to bits of broken-off coral. It has an intensive yellow-orange colour on the inside of the pectoral fins, as has the devilfish. This is displayed as a warning signal by turning the pectoral fins (see photo page 119).

Found: In tropical and temperate seas, some species in cold seas also. They prefer firm sea bed and are common in coral and rock reefs.

Way of life: Scorpionfishes are poor swimmers and only swim short distances, settling back on the ground for a longer period of time. They lie in wait for prey for hours without moving, until an unsuspecting victim swims right up to their mouths. They open their mouths rapidly, creating a current which sucks in the prey. They can overpower astonishingly big fish. When they change places they can adjust their colour in seconds.

Food: Fish, but also crustaceans.

Reproduction: They lay several thousand eggs in gelatine-like clumps.

Warning: Scorpionfish are equipped with many poisonous spines, which are used only in defence. However, they are still dangerous to divers, as they rely on their good camouflage. The front part of the dorsal fin has 12-14 poison spines, the pelvic fin one and the anal fin three. Poisoning causes severe pain and possible paralysis, but is not believed to be dangerous to human life.

Leaf fish, *Taenianotus triacanthus*; up to 8 cm. When in danger it falls onto its side and imitates a floating leaf. Colour varies from yellow and pink to black. East Africa to the Galapagos. Photo: Maldives.

Tasselled scorpionfish, *Scorpaenopsis oxycephalas*; up to 30 cm. The scorpionfishes are hard to identify. Red Sea to west Pacific.

Devil scorpionfish or false stonefish, *Scorpaenopsis diabolus*; up to 22 cm. This species is often taken for a stonefish by divers. Red Sea to Tahiti. Photo: Thailand.

Stone- and Devilfishes – Synanceiinae and Choridactylinae

Sub-families Synanceiinae and Choridactilinae, Family Scorpaenidae, Order Scorpaeniformes. Few species. Size: stonefishes up to 60 cm, devilfishes up to 18 cm.

Characteristics: Stonefishes are among the best camouflaged of all fishes. Their plump bodies taper sharply towards the rear, so that their huge heads look out of proportion. The large mouth points almost vertically upwards, the eyes are set very high. The front of the dorsal fin is composed of 12-14 sturdy poison spines which are covered with a fleshy skin. The extremely large, fleshy pectoral fins have rounded serrations and are the first thing a practised observer will notice. The entire body surface is covered with irregular warts and rags of skin. Camouflage is perfected by colouring and an asymmetrical body position in which the tail is curled to one side in a hooked shape.

Devilfish differ from stonefish in their smaller size and more delicate body. The profile of the head is concave; only the eyes are raised right up above all else on bony ridges. The dorsal fin consists of 15-17 long poisonous spines which are asymmetrically laid to either side and camouflaged with rags of skin. The two bottom rays of the pectoral fins are not connected by membrane; they are used to propel the fish for crawling. The tail fin is folded flat and, like the pectoral fins, brightly coloured with warning patterns and colours. If you approach a devilfish it will open out the tail fin like a fan and turn up its pectoral fins.

Found: From the Red Sea to the Pacific on sandy or stony beds in water that is not too deep; often hidden in caves or crevices during the day.

Way of life: Stonefishes are poor swimmers and do not like changing positions. Their slow movements would hardly be enough to make them swim if the large pectoral fins did not have "sail-like" qualities. Their swimming skills are not sufficient for hunting prey. Stonefishes are obliged to wait until a fish swims close enough to their mouths to be sucked in. They swallow fish up to their own size.

Food: Fish and crustaceans.

Warning: The poison of stone and devilfishes is extremely painful and can in some cases be deadly.

Stonefish, *Synanceia verrucosa*; up to 38 cm. Likes to hide in caves and crevices by day. The curled tail is typical for stonefish. Red Sea to Tahiti. Photo: Egypt.

Estuarine stonefish, *Synanceia horrida*; up to 60 cm. Sometimes overgrown with algae. Prefers sandy and muddy bottom, often in brackish water. Thailand to Australia. Photo: Thailand.

Spiny devilfish, *Inimicus filamentosus*; up to 18 cm. When in danger the intensively coloured tail and pectoral fins are displayed (see photo on right). Red Sea to Indonesia. Photo: Egypt.

Lionfishes – Pteroinae

Sub-family Pteroinae, Family Scorpaenidae, Order Scorpaeniformes. 11 species. Length: 15-35 cm.

Characteristics: Lionfishes are conspicuous in shape, colouring and patterning. Many have feather-like feelers which look like horns between the eyes. These are shed when the fish become adult. The young, up to a size of about 5 cm, are colourless. Males have a more powerful head. The males become darker in shade during mating.

Found: From the Red Sea to the Pacific, at depths of up to 40m.

Way of life: All lionfishes are slow, majestic swimmers, and mostly stay close to the bottom. They are very rarely seen in the open water above a reef. In certain circumstances they can shoot forward for a short distance much faster than one would expect. However, they do not have the necessary staying power to chase their prey. They can be seen singly as well as in groups of up to 12. The males live in territories, which may however overlap. Lionfishes are particularly active at dusk and have developed a specialised hunting technique. They use the extremely elongated rays of their fins rather like a net, driving potential victims into a favourable position, such as a niche, in order to suck them up with lightning speed. The poisonous spines on their dorsal fins are evidently not used for hunting but serve as a defence.

Food: Fish.

Reproduction: During mating, combat between rivals is not confined to males. Females are also attacked by the extremely aggressive males. During this time it can also be dangerous for divers to get too close to the "bridegroom". Mating and spawning take place shortly after sunset. The couple swim up into the open water. When they have reached the highest point the eggs are released and drift away with the current.

Lionfish, *Pterois volitans*; up to 35 cm. According to the most recent studies, the similar species *P. miles* is confined to the Indian Ocean. Thailand to Tahiti. Photo: Thailand.

Zebra lionfish, *Dendrochirus zebra*; up to 25 cm. The pectoral fins are linked with membranes. East Africa to Samoa. Photo: Thailand.

Warning: the placid behaviour of lionfishes tempts many divers to approach them. If the approach is close enough for the fish to want to flee, it will usually back away. Should this not be possible, however, the fish could make a direct attack. It will shoot forward at lightning speed and point its poisonous spines at its supposed attacker. Stings can be extremely painful and cause large-scale swelling and paralysis.

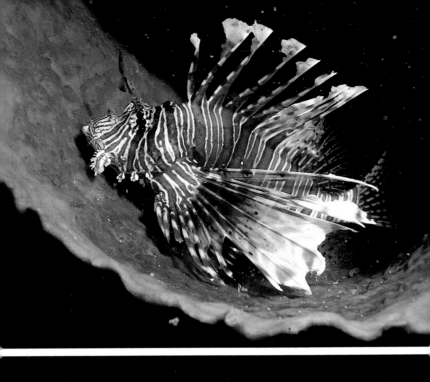

Lionfishes – Pteroinae

1. **Spot-fin lionfish**, *Pterois antennata*; 22 cm. Red stripes on the tail run diagonally. Way of life: lagoons and outer reefs from reef platform to depths of 50 m. Hides in crevices or caves during the day. Hunts at night, catching crustaceans. East Africa to Marquesas.

2. **Radial lionfish**, *Pterois radiata*; 23 cm. Stripes at the base of the tail run parallel. Way of life: shallow lagoons and outer reefs to depths of 20 m, sometimes sharing a hiding place with the spot-fin lionfish. Red Sea to Society Islands.

3. **Mombasa lionfish**, *Pterois mombasae*; 16 cm. Markedly spotted pectoral fin, dark spot near the head. Way of life: generally near coral reefs below 40 m. Rare. Southern Africa, Sri Lanka and New Guinea.

4. **Pacific devil lionfish**, *Pterois volitans*; 38 cm. Similar to P. miles of the Indian Ocean, which has 13 dorsal spines as against *P. volitans'* 11. They are almost impossible to distinguish underwater. Way of life: common in lagoons and on outer reefs up to depths of 50 m. Spends the day under overhangs, hunts crustaceans and small fish at dusk and at night. Sometimes approaches divers. Western Australia to Marquesas; *P. miles* from the Red Sea to the eastern Indian Ocean.

5. **Japanese lionfish**, *Pterois lunulata*; 30 cm. No, or few, spots on the hind fins. Way of life: outer reefs below 20 m. Rare. Southern Africa to western Pacific.

6. **Hawaiian lionfish**, *Pterois sphex*; 22 cm. Pectoral spines reach the base of the tail. Dark banded body. Way of life: among rocks and coral reefs at depths from 3 to 13 m. Hides by day under overhangs and in caves. Nocturnal. Endemic in Hawaii.

7. **Zebra lionfish**, *Dendrochirus zebra*; 20 cm. T-pattern at the base of the tail. Pectoral fins have membranes. Way of life: shallow coral reefs. Males are aggressive and drive other males out of their territory. East Africa to Samoa.

8. **Twinspot lionfish**, *Dendrochirus biocellatus*; 10 cm. Two characteristic "eye" spots on the dorsal fins. Way of life: lives in clear water near exposed coral-rich reefs at depths between 1 m and at least 40 m. Sri Lanka to Society Islands.

9. **Short-fin lionfish**, *Dendrochirus brachypterus*; 18 cm. The rays of the pectoral fins are linked by a membrane. Way of life: prefers isolated rocks with algae growths in sandy parts of the reef platform or in lagoons. Often assumes a "head over heels" position in hiding. Nocturnal. Red Sea to Samoa.

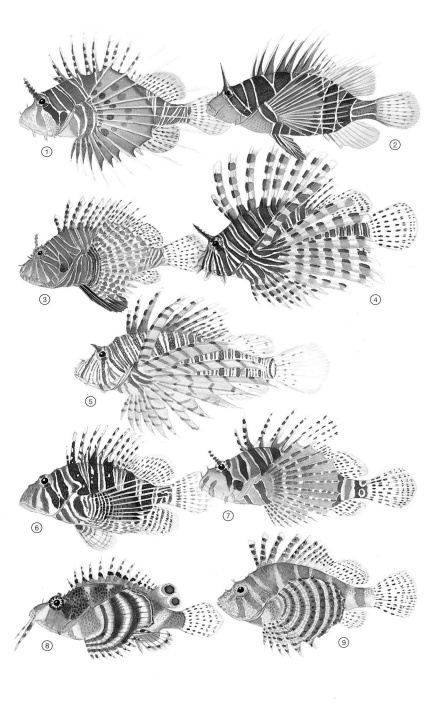

Basslets – Anthiinae

Sub-family Anthiinae, Family Serranidae, Order Perciformes. At least 40 species. Size: up to 12cm.

Characteristics: Vertically compressed body with typical bass-like form. The head is pointed, the mouth at the end is small; the fins are relatively large and pointed. The tail fin is deeply forked or crescent-shaped. The dorsal fin is continuous along the back but may be separated. In males the second or third dorsal ray is elongated, extremely so in some species. The body is covered with clearly visible scales. Almost all species are colourful, but conspicuous patterning is rare. Species with three or four colours are often seen. The colours generally fade into one another like water-colours. The basic colours are usually red, orange, yellow or violet. Males and females are differently coloured.

Found: In tropical and temperate seas from the Red Sea to the Pacific (also in the Mediterranean below 30 m). They spend their time in open water above reefs and when in danger or at night look for protection in the reef. There are more of them to be seen in coral reefs that in rocky areas.

Way of life: Basslets are very lively swimmers with good staying power. They swim in large shoals of hundreds or even thousands above shallow, sheltered coral reefs or hunt for plankton along current-rich outer reefs. These diurnal creatures never seem to tire, not even in a strong current. They are always on the move, except when allowing a cleaner fish to remove their parasites. Young adults are all female and live in a "harem". There are at least five females to a male, many more are possible. If a male dies, the female of highest rank changes into a male. (According to Fishelson, the presence of males blocks the change of the females). Males are territorial and slightly bigger than females. Their more splendid colouring is noticeable, and many have an elongated dorsal fin ray.

Food: Exclusively plankton-eaters.

Reproduction: The lyretail fairy basslet, *Pseudanthias* (or *Anthias*) *squamipinnis*, seems to spawn only in the winter months in the Red Sea, while the same species along the Great Barrier Reef only spawns in the hot summer months. Courtship takes place at dusk, with the male - followed by the female - performing zig-zag dances.

Lyretail fairy basslet, male (see also centre photograph). Photo: Egypt.

Lyretail fairy basslet, *Pseudanthias squamipinnis*; up to 12 cm. Female. A male has a harem of 6-8 females; they usually appear in large schools. Red Sea to Solomon Islands. Photo: Thailand.

Yellowtail basslet, *Pseudanthias evansi*; up to 10 cm. Male: females are less intensive in colour. Kenya to Thailand. Photo: Thailand.

Groupers – Epinephelinae

Sub-family Epinephelinae, Family Serranidae, Order Perciformes. They are the largest bony fishes and can reach lengths of 3.5 m and a weight of 550 kg. However, there are also small species which only reach 25 cm. The sub-family comprises some 350 species.

Characteristics: Very robust fishes with a powerful, slightly laterally compressed body. The large head and huge mouth are conspicuous. The lower jaw protrudes noticeably. The large, protuberant eyes have an egg-shaped pupil which often has a yellow border.

The long dorsal fin has strong spines in the front. Their number varies from 7 to 11 according to genus. When at rest the large tail fin is folded relatively small. The shape of the tail fin varies greatly from round, straight, slightly convex or concave to crescent-shaped. Pectoral and pelvic fins are always set close above each other; their first few rays often develop into powerful spines. Many species match the sea bed in their colouring. The patterning almost always consists of variously formed large blotches which camouflage the fish well; blurred bands across the body are also frequent. The groupers, which often appear red in artificial light, can only be see with great difficulty in their natural surroundings, due to the filter effect of the water in the depths.

Found: In all tropical and temperate seas near to the coast. They live hidden in the reefs near the sea bed and occur at depths of up to 200 m; they prefer rocky shores.

Way of life: Groupers are predators and most of the time they wait, well camouflaged, for their prey. They are solitary creatures and defend their territories energetically against others of their species. The older they become, the more their territory expands, and they have sites which they prefer such as caves, crevices and corals. Only the young sometimes live in groups. They do not as yet have a strong sense of territory.

When swimming through their territory groupers are very calm and relaxed. They can speed up their swimming and reach quite unexpected speeds. Large groupers have become timid due to harpooning.

Duskyfin grouper, *Cephalopholis urodeta nigripinis*; up to 25 cm. In this picture displaying warning colours. Prefers coral-rich reefs. Indian Ocean. Photo: Thailand.

Peacock grouper, *Cephalopholis argus*; up to 50 cm. In many regions this is the most common grouper. Red Sea to central Pacific. Photo: Maldives.

Coral grouper, *Cephalopholis miniatus*; up to 36 cm. Popular with photographers. Prefers coral reefs. Red Sea to central Pacific. Photo: Maldives.

Groupers (continued)

Species of the genus *Cephalopholis* are very common in tropical areas and known as coral groupers. They can change colour swiftly. The order Serranidae has some of the most expensive food fishes among its members. They are almost all caught only by angling. All Serranidae, including the groupers, can change sex. They begin as females but can change into males.

Food: Fish and crustaceans. Groupers can swallow very large chunks which they regurgitate if food becomes stuck in their throats. They have many very tiny teeth which are needle-sharp and point backwards. These are useful for catching prey but not suitable for breaking up large chunks of food. When catching prey the large mouth creates a current which sucks in prey. Large groupers probably have proportionally the largest mouths in the entire animal kingdom.

Reproduction: Groupers spawn in open water at certain times of the year and phases of the moon. The larval stage lasts for several weeks. In some areas certain species travel several kilometres to reach their spawning grounds. In the Bahamas the number of individuals in the spawning schools has been estimated at 100,000. The eggs drift at low tide out from outer reefs and reef channels to the open sea. There the larvae are relatively safe from predators. When the larval stage is over the young return to the reef area. There are also species that travel to fresh water to spawn and lay up to a million eggs there. The young spend the first year of their lives in the rivers in which they have hatched.

Warning: Fully grown groupers can be very poisonous due to ciguatera, because as predators they are at the end of the food chain. It is said that certain fish of the largest species, the giant grouper *Epinephelus lanceotatus*, have swallowed pearl divers. Feeding groupers is not always without its dangers either. It is easy to forget that they too get bigger and lose their fear of humans. I have personally experienced this with a hand-fed grouper, only 60 cm long – it suddenly had my hand in its mouth.

Lyretail grouper, *Variola louti*; up to 80 cm. The basic colour varies from violet to reddish-brown. The very similar species *V. albimarginatus* has a white border at the end of its tail. Red Sea to Tahiti. Photo: Egypt.

Malabar grouper, *Epinephelus malabaricus*; up to 2 m. Can be identified by the little dark spots. Red Sea to Sri Lanka. Photo: Egypt.

Leopard grouper, *Plectropomus pessuliferus marisrubri*; up to 80 cm. Basic colour from red to brown. Can only be distinguished from *Epinephelus leopardus* and *E. laevis* by the dark-edged blue spots. Red Sea to Fiji. Photo: Egypt.

Hawkfishes – Cirrhitidae

Family Cirrhitidae, Order Perciformes. Small species, mostly under 10 cm in length. A few of the 24 species may reach 20 cm and more.

Characteristics: Hawkfishes have a perch-like form with a laterally compressed body. They have only one dorsal fin with 10 dorsal fin spines in the front part; the membranes are deeply indented. At the tips of the spines there are little bushes of small hair-like attachments. The pectoral fins are also deeply indented between the rays and are spread like fingers. The creatures can get a good purchase on the bottom by sticking their fin spines into holes and cracks. Hawkfishes have scales and possess a continuous lateral line. Their colour and patterning vary greatly. The shape of the head also varies from blunt and rounded to an elongated pointed snout.

Found: Hawkfishes are found in all tropical seas. They inhabit reefs that are not too deep and are usually found in higher spots, with the exception of the long-nosed hawkfish, *Oxycirrhitus typus*. Hawkfishes live only above firm sea bed, coral as well as rocks.

Way of life: The fish have earned their name "hawkfish" by their typical behaviour. From exposed points they watch over their territories as if on sentry duty, hovering motionless with rolling eyes. They have a hiding place nearby to which they can retreat when endangered. Hawkfishes cannot swim for long and can only manage short stretches. They have little capacity to pursue fleeing prey. These diurnal fish are always seen singly, never in groups. Hawkfishes, like the Serranidae previously described, are hermaphroditic, female in youth but able to turn into males when needed. It is not possible for them to reverse the process.

Food: Crustaceans and small fish.

Reproduction: A male always has several females which form a "harem". Hawkfishes only spawn at night or at dusk, swimming rapidly to the surface in pairs and ejecting the eggs into the open water. The larval stage lasts for several weeks.

Pixie hawkfish, *Cirrhitichthys oxycephalus*; up to 8 cm. Colour and pattern vary. Red Sea to eastern Pacific. Photo: Thailand.

Arc-eye hawkfish, *Paracirrhites arcatus*; up to 13 cm. The eye ring is a typical feature. Prefers stag's-horn coral. East Africa to Hawaii, Tahiti. Photo: Maldives.

Blackside hawkfish, *Paracirrhites forsteri*; up to 20 cm. This species can change colour according to background. Red Sea to Hawaii, Tahiti. Photo: Thailand.

Cardinal fishes – Apogonidae

Family Apogonidae, Order Perciformes. Most species about 6 cm in length, only a few exceed 15 cm. There are about 190 species.

Characteristics: The laterally compressed body varies in height between species. Some are relatively slender, others seem short or squat. Males have a larger head than females. The large eyes are another feature. Both dorsal fins are high and clearly separated; they are often triangular in shape. Tail fin shapes vary greatly from rounded to straight to forked. The base of the tail is notably far forward. Many species have horizontal stripes and are inconspicuously coloured. The bright red species which have given the family its name are rare. The scales are large and usually clearly visible.

Found: In all tropical and temperate seas, even at great depth. They are most common in the Indo-Pacific region. Some species live in brackish water. They can be found in a variety of corals: the smaller species prefer horny corals, while the larger species look for protection between corals or under overhangs.

Way of life: Cardinal fishes are not very fast swimmers. They live on a particular site in large groups, but do not – as is common with most shoals of fish – patrol a certain territory, but hover between the branches of coral or in caves. Not until dusk or nightfall do they spread out to hunt. Their space of action is extremely small for a fish. There are species that live amid the spines of a sea urchin, with crown of thorns starfish or in rare cases even in the tentacles of a sea anemone, and they are limited to that tiny space. If they venture too far out of this protected area they run the risk of being eaten.

Food: Zooplankton, crustaceans and small fish.

Reproduction: Most species, if not all, hatch eggs in their mouths. The eggs – about 100 – are placed in the male's mouth until the young hatch. This is probably why the male's head is larger than the female's. After hatching the young always stay close enough to the parent to seek protection in its mouth when danger threatens.

Split-banded cardinal fish, *Apogon compressus*; 11.5 cm. Typical feature is the brown line of dots at the base of the dorsal fin. Thailand to west Pacific. Photo: Thailand.

Five-lined cardinal fish, *Cheilodipterus quinquelineatus*; up to 32 cm. Red Sea to Tahiti. Photo: Thailand.

Big-eyes – Priacanthidae

Family Priacanthidae, Order Perciformes. About 18 species. Size: up to 35 cm long.

Characteristics: Laterally compressed body: head profile more strongly convex below than above. Lower jaw protrudes markedly, the mouth points diagonally upwards. They have extremely large eyes. The dorsal fin is continuous and has 10 spines in the front part. Tail fins can be convex as well as concave. Most species are reddish to bright red in colour; at night they are silvery. This colour change can sometimes also be seen by day. Some species also have a pattern of spots.

Found: World-wide in tropical or temperate seas, at depths of 2-200 m. Under overhangs by day.

Way of life: These nocturnal fish can also be observed by day. During the day they hover in caves or under corals and hardly move. Not until nightfall do they look for food. They are usually seen in small groups - also singly.

Food: Large zooplankton.

Reproduction: Long larval stage in open water.

Suckers – Echeneididae

Family Echineididae, Order Perciformes. Eight species. Size: maximum length 90 cm.

Characteristics: Round, elongated body with elongated sucking disk on the head which is not unlike a deep-treaded sole; it has developed from the first dorsal fin. The tail fin is well developed. The lower jaw protrudes far beyond the upper. Adult fish are often dark brown above and on the sides, and the belly is yellowish. Most species have pale horizontal stripes.

Found: World-wide, but more common in warmer areas. They follow whales and large fish into the open sea. In reef areas they are to be seen attached to smaller fish or turtles or swimming freely without a host animal.

Way of life: Suckers have the ability to attach themselves to other creatures by their sucking pads and be pulled along by them. However, they are not parasites, as many work as cleaner fishes and keep their hosts clear of parasites. Some species have specialised in one sort of host; others change from one species to another. It is not unusual for them to cling to humans by mistake.

Food: Waste and parasites off its host animal.

Goggle-eye, *Priacanthus hamrur*; up to 32 cm. This species is mostly dark red by day, silver at night. The colour change takes place in seconds. Photo: Thailand.

Glass-eye, *Heteropriacanthus cruentatus*; up to 28 cm. Typical feature is the tail fin with its straight "chopped-off" appearance. Photo: Thailand.

Shark sucker, *Echeneis naucrates*; up to 1 m. This one has attached itself onto a zebra shark. All seas. Photo: Thailand.

Carangids – Carangidae

Family Carangidae, Order Perciformes. More than 200 species. Size: from 20 cm to almost 2 m in length; large species have become rare.

Characteristics: The body is laterally compressed and usually relatively deep. The young fish have deeper bodies than the adults. Carangids have scales that vary in size and a remarkably slender tail base. The tail fin is very high and deeply forked. The first of the two dorsal fins is rarely seen because it can be retracted into a groove, making its tip lie flush with the surface of the body so that the join is hardly visible. The second dorsal fin is almost symmetrically opposite to the anal fin; it has very long spines in the front part and then forms a short fringe leading to the base of the tail. This gives the fish its typical shape (see also diagram). The long, slender, sickle-shaped pectoral fins are also very marked, but as they are usually transparent they are hard to recognise. The continuous lateral line ends in a wedge-like, bony reinforcement of the tail base. Carangids differ in colour but most species are silvery, with a metallic blue or green back; the fins are often yellowish. Juveniles can also be yellow along the sides or have dark bands across the body.

Found: World-wide in tropical and temperate seas. Carangids prefer to swim near the surface. They roam in the open sea and sometimes hunt in the vicinity of reefs. There are, however, also species which always live in reefs.

Way of life: As inhabitants of the open sea carangids are swift and enduring swimmers and can reach up to 50 km/h. Generally they can be observed in shoals of varying sizes, and are occasionally met individually. They are active by day and night and constantly in motion, otherwise they would sink; the swim bladder is atrophied or missing. When hunting they try and break up shoals of fish by swimming into the shoal at high speed from different directions. Fish which hesitate for a moment while deciding which part of the shoal to follow are potential victims. Juveniles are often to be seen under the protection of jellyfish hunting for plankton. When in danger they hide among the stinging tentacles (see also pages 42/43). Carangids are important for commercial and private fishing.

Food: Fish, crustaceans and other invertebrates.

Reproduction: Largely unknown as yet; presumably pelagic, in large schools.

Bluefin travelly, *Caranx melampygus*; up to 80 cm. Fast swimmers from the open sea, which often hunt along the slopes of reefs. Red Sea to eastern Pacific. Photo: Maldives.

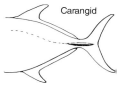
Carangid

Yellow spotted travelly, *Carangoides bajad*; up to 53 cm. the same species can be silvery with yellow spots. Red Sea to Indonesia. Photo: Egypt.

Snappers – Lutjanidae

Family Lutjanidae, Order Perciformes. Four sub-families, 103 species. Size: up to 1 m, most species less than 30 cm.

Characteristics: Robust fishes with laterally compressed bodies which can vary in depth. Some are slender, others could be described as squat. In the genus *Lutjanus* the head is almost triangular with an elongated snout. The genus *Macolor*, in contrast, has a rounded blunt head. The mouth is at the end and can be protruded; it has conspicuously large teeth. The eyes are relatively large, the body covered with scales. The dorsal fin is continuous, high at the front, with spines, curving back in the softer-spined rear. The tail fin is well developed, concave or slightly forked in shape. In juveniles it can be round or convex. The tail base is relatively deep. Colour and patterning vary; most fish are uniform in colour or have horizontal stripes, often with a dark spot as well. The juveniles of the genus *Macolor* are remarkable for their black and white colouring.

Found: In tropical and sub-tropical seas. They live protected by the reefs, under overhangs, among the corals, in caves, along the outer reefs, but also in schools in the open water above the reefs. They generally live at depths of 2-40 m, red-coloured species up to 80 m. *Lutjanus kasmira* has even been found at 265 m in the Red Sea. Some species live in brackish water.

Way of life: Snappers are not active swimmers and do not swim great distances. Many species can be found in very large schools. They are partly diurnal, partly nocturnal. They do not lie on the sea bed like the Serranidae, but hover, often pressed close together, near corals or above sandy bottom. Other species shoal in the breaker zone or are solitary. They remain near their specific sites. This family is very important for commercial fishing, as the flesh is highly prized.

Food: Small fish, crustaceans and plankton.

Reproduction: Little is known, except that the blue-lined snapper, *Lutjanus kasmira*, mates at dusk. The male encourages the female to release eggs by nudging her belly. The female swims upwards in a spiral towards the surface and releases the eggs into open water.

Blue-lined snapper, *Lutjanus kasmira*; up to 30 cm. The most common species. The fish spend the day close together in the protection of the reef. Red Sea to Tahiti. Photo: Maldives.

Black-spot snapper, *Lutjanus ehrenbergi*; up to 30 cm. East Africa to Samoa. Photo: Egypt.

Twinspot snapper, *Lutjanus bohar*, up to 90 cm. Fully grown fish lose the white signal spots. Red Sea to west Pacific. Photo: Thailand.

Fusiliers – Caesionidae

Family Caesionidae, Order Perciformes. They are related to the snappers and were once biologically grouped with them. Some 25 species. Size: up to 30 cm.

Characteristics: Slender, spindle-shaped body, small head with large eyes and mouth at end. The long continuous dorsal fin generally lies close to the body so only the well-developed and deeply forked tail fin is noticeable. Many species are two-coloured, some even three-coloured. Blue, yellow and turquoise are the most common colours. They can change colour to red at night. Patterns vary greatly. Horizontal stripes and bands are most common. Some fish have two dark spots at the ends of the forked tail fin. The diagonal line separating blue and yellow in the yellowback fusilier *Caesio teres* is very conspicuous. Fusiliers have small scales; the clearly visible lateral line is continuous.

Found: In tropical seas, mainly in open water. They occasionally swim through reefs at medium depths.

Way of life: Fusiliers are fast, enduring swimmers and are in motion all day long. They only allow themselves a short break when they visit a cleaner fish. Often several are waiting until it is their turn. If the shoal moves on they follow, without being freed of their parasites. Fusiliers only occur in shoals, which can be very large. They are well adjusted to life in open water, where they hunt plankton. Here they are, however, often hunted by large predatory fish. They are relatively safe in a large shoal, as it is difficult for an attack to fix on a single fish in the dense milling crowd. A shoal is not always composed of only one species; two or three species may live together. At night fusiliers hide in the reefs under overhangs and in crevices.

I would like to tell you one of my personal experiences to illustrate how large shoal of fusiliers can be, and what an extraordinary effect they can have on a diver. I had glanced in the direction of the open sea, hoping to see some large fishes, when the water turned yellow within a few seconds. I was most taken aback until I noticed that a huge shoal of yellow fusiliers in flight was coming towards me at high speed, looking for protection in the reef. There were some two thousand fish in the shoal.

Food: Exclusively zooplankton.

Reproduction: Largely unknown.

Yellowback fusilier, Caesio teres; up to 30 cm. They hunt zooplankton by day. East Africa to Samoa. Photo: Maldives.

Bluestreak fusilier, Pterocaesio tile; up to 25 cm. They lose the red colouring in open water. East Africa to Tahiti. Photo: Maldives.

Sweetlips (grunts) – Haemulidae

Family Haemulidae, Order Perciformes. Also known as grunts. Right up to the present day it is still not clear which name is to be preferred, so I have decided on "sweetlips", which is more widely known in diving circles. The family comprises 175 species in 17 genera. They grow up to a length of 27-95 cm.

Characteristics: Sweetlips have a form like other Serranidae but are distinguished by their pouting lips. Their bodies are laterally compressed and often relatively deep. They have a long, high dorsal fin, the first part of which consists of 9-14 spines linked by membrane. The tail fin is well developed and takes various forms: straight, convex, concave, or slightly forked. Juveniles mostly have rounded tail fins. Sweetlips are generally magnificently coloured fishes with bright "boiled sweet" colours – hence the name. They undergo great changes as they grow, not just in colour but also in pattern. Many juveniles have horizontal stripes or spots and are hard to identify. Others are brightly coloured with large spots. When they are fully grown the conspicuous colouring usually goes and the fish remains finely spotted or striped.

Found: In all tropical seas, among reefs that are not too deep and close to the shore. Only a few species are to be found in temperate seas.

Way of life: Sweetlips often occur in schools of varying sizes. In some areas they may be several hundred in one school. These nocturnal fish usually spend the day close together under corals. The juveniles differ from adults not only in their external appearance. They are only seen singly. Their compulsion to move has nothing in common with the behaviour of adult sweetlips. They make rapid serpentine movements while swimming without moving forward. They follow no particular direction but swim up and down apparently aimlessly within a small territory, never keeping still for a second. It remains to be proved whether this behaviour represents mimicry of inedible tropical nudibranchs. The name "grunt" comes from the ability some species have of making noises, which are amplified by the swim bladder, by grinding their pharyngeal teeth. Sweetlips are not very timid and in some areas have been greatly decimated by harpooning.

Food: Bottom-dwelling invertebrates.

Harlequin sweetlip, juvenile (see central picture). Photo: Thailand.

Harlequin sweetlip, *Plectorhinchus chaetodonoides*; up to 70 cm. Adult colouring (for juvenile see above). Mauritius to New Caledonia. Photo: Maldives.

Oriental sweetlip, *Plectorhinchus orientalis*; up to 60 cm. Common species in reefs. Juveniles are dark brown with cream-coloured spots. East Africa to Samoa. Photo: Maldives.

Sweetlips – Haemulidae

1. **Oriental sweetlip**, *Plectorhinchus orientalis*; 80 cm. Black horizontal stripes. Juveniles resemble nudibranchs. Way of life: common species, which spends the day in small groups under coral. Prefers outer reefs with thick coral growth. Depth – 2-25 m. East Africa to Samoa.

2. **Lined sweetlip**, *Plectorhinchus lessoni*; 45 cm. No stripes on the belly. Way of life: hides by day under coral protuberances in sheltered lagoon reefs and outer reefs up to depths of at least 25 m. East Africa to New Caledonia.

3. **Diagonal sweetlip**, *Plectorhinchus lineatus*; 70 cm. Black stripes are diagonal, no stripes on the belly. Way of life: this timid species spends the daylight in small groups in coral-rich outer reefs or reef channels at depths of 2-30 m. West Pacific, Ryukyus, southwards to Great Barrier Reef.

4. **Black-spotted sweetlip**, *Plectorhinchus gaterinus*; 60 cm. Juveniles have six black stripes on the sides which fade gradually. Way of life: in groups in coral reefs, especially in outer reefs, to a depth of 30 m. Red Sea to southern Africa.

5. **Two-lined sweetlip**, *Plectorhinchus albovittatus*; 40 cm. Adults similar to juveniles but lower lateral stripe fades. Way of life: solitary on coral-rich reefs up to depths of 30 m. Red Sea to Celebes.

6. **Many-striped sweetlip**, *Plectorhinchus polytaenia*; 40 cm. Typical yellow-brown lateral stripes. Fins are yellow. Way of life: spends the daylight in sheltered places on coral or rock reefs. Indonesia to Western Australia, India.

7. **Celebes sweetlip**, *Plectorhinchus celebecus*; 45 cm. Slender yellow lateral stripes on grey base. Fins are yellow. Way of life: solitary in sheltered reefs at depths of 8-25 m, in the vicinity of coral outgrowths and isolated coral trees. Ryukyus to southern Great Barrier Reef.

8. **Lemon sweetlip**, *Plectorhinchus flavomaculatus*; 60 cm. Yellow laterals stripes largely disappear as the fish ages except for the head. Way of life: in rock and coral reefs, also in areas where algae grow. Red Sea to western Pacific.

9. **Goldspotted sweetlip**, *Plectorhinchus multivittatum*; 40 cm. Yellow stripes dissolve into spots. Yellow fins. Way of life: little is known. Solitary in coral reefs. Eastern Indian Ocean and western Pacific.

10 **Spotted sweetlip**, *Plectorhinchus picus*; 85 cm. Dark spots on pale background. Juveniles are black and white. Way of life: inhabits lagoons and outer reefs at depths of 3-50m. Solitary by day under coral overhangs, Seychelles to Society Islands.

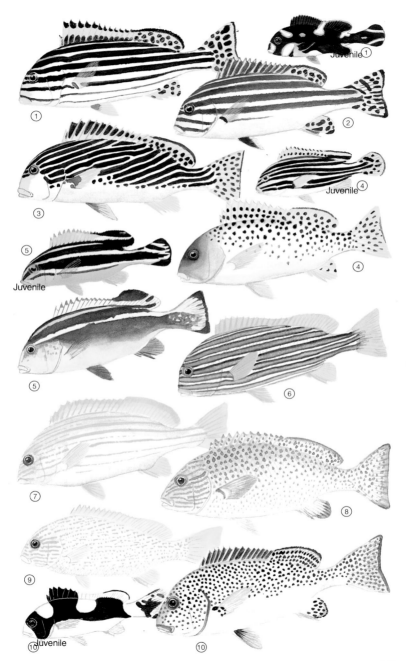

1

Juvenile 1

2

3

Juvenile 4

5

4

Juvenile

5

6

7

8

9

10 Juvenile

10

Nemipterids – Nemipteridae

Family Nemipteridae, Order Perciformes. Numerous species. Size: up to 40 cm long.

Characteristics: Nemipterids are slender, laterally compressed with a small head, large eyes and a short snout; the mouth is small. They have a continuous dorsal fin which has 10 spines in the front part. The well developed tail fin is slightly forked; the anal fin has three slender spines. Nearly all species have large scales which cover the whole body but not the head. Many of them have marked patterns, others are a delicate pastel colour. The juveniles can look very different from the adults. The genus *Scolopsis* has a characteristic feature - just below the eye there is a powerful backward-pointing spine which gives this genus the name "spinecheek". Sex change has been observed in some species. It is assumed that all species of the family start off as females and may then transform into males. The change is controlled by "social factors".

Found: In the Red Sea, Indian Ocean to the Pacific. They usually live in reefs and shallow water in coastal regions.

Way of life: Nemipteridae are neither fast nor lively swimmers. They often hover motionless between corals, then swim a few metres, then stop motionless in one spot again. This behaviour is characteristic of the family. They live individually or in small groups and are rather inconspicuous among the brightly coloured life of a coral reef. Many species prefer sandy places with single outcrops of coral in sheltered coastal waters. In some areas they are fished for food.

Food: Bottom-dwelling invertebrates such as bristle worms or small crustaceans.

Reproduction: Little is known; mating probably takes place at night in large schools outside the reefs. Juveniles start off female and may later turn into males.

Brown-striped spinecheek, *Scolopsis affinis*; up to 25 cm. Juvenile; this could also possibly be a species as yet unclassified. India to Indonesia. Photo: Thailand.

Ciliated spinecheek, *Scolopsis ciliatus*; up to 20 cm. the white dorsal stripe shows up brightly under water. India to Micronesia. Photo: Thailand.

Emperors – Lethrinidae

Family Lethrinidae, Order Perciformes. They are closely related to the snappers. There are only 20 species. Size: up to 90 cm.

Characteristics: All species have laterally compressed bodies and large scales except on the snout. This differentiates them clearly from the snappers. The very high position of the eyes is notable in most species. The continuous dorsal fin has 10 spines in the front part: the anal fin has three. All other features vary considerably, especially the shape of the head. Some species have a blunt head, others a pointed mouth; the most remarkable species, *Lethrinus elongatus*, has an extremely elongated concave snout. Many species have thick lips, a medium-sized mouth and relatively large eyes. Colouring is usually inconspicuous; patterns vary from horizontal to vertical stripes to irregular blurred patches; some are uniform in colour. In some species marked colour changes can be observed which take place during the mating season or at night.

Found: In all tropical seas.

Way of life: Their behaviour varies: while the bigeye emperor, *Monotaxis grandoculis*, spends the daylight hours solitary or in small groups in the open water above the reef slopes, species belonging to the genus *Lethrinus* swim over sandy sea beds in the vicinity of reefs, looking for food. The gold-spotted emperor, *Gnathodontex aurolineatus*, is always to be found inactive in dense schools between corals. Most species are nocturnal. All species can change their sex; juveniles are female and can become male. The process cannot be reversed.

Food: Hard-shelled creatures and fish.

Reproduction: Largely unknown.

Warning: Ciguatera poisoning cannot be ruled out with some larger species if the fish are eaten.

Bigeye emperor, *Monotaxis grandoculis*; up to 55 cm. Swims above the reef slopes, solitary or in small groups. Red Sea to Hawaii. Photo: Egypt.

Longnose emperor, *Lethrinus olivaceus*, previously known as *L. elongatus*; up to 1 m. Mostly moves in small groups, looking for food, among blocks of coral, digging up the sea bed. Red Sea to Samoa. Photo: Thailand.

Orange-fin emperor, *Lethrinus erythracanthus*; up to 80 cm. Timid solitary species. Previously known as *L. kallopterus*. East Africa to Samoa. Photo: Maldives.

Goatfishes – Mullidae

Family Mullidae, Order Perciformes. About 60 species. Size: up to 55 cm. Females grow larger than males.

Characteristics: Goatfishes have a laterally compressed, longish body which is notable for its profile. The head and back are more convex than the underside of the fish. The eyes are set very high and far back. The most conspicuous features are the two long barbels which are under the chin and point backwards when swimming. They cannot then be seen, as they lie in grooves. These organs of touch are equipped with taste buds. When looking for food they are swung forward and very skilfully prodded into all the holes and cracks to track down prey. The mouth, set at the end, has thick lips which can be everted. The first of the two relatively high dorsal fins is mostly laid flat against the body. The well developed tail fin is deeply forked, the body covered with large, easily visible scales. The colour of the fishes varies across a spectrum of red, yellow, blue to pale colours with the greatest variety of patterns.

Found: Widespread in tropical and temperate seas world-wide; generally live over sandy bottom, mud or sea beds with broken coral.

Way of life: Goatfish can be diurnal or nocturnal or, according to species, both. They often swim above the sea bed in groups looking for food. When using their barbels to track down food on the sea bed, they will grub up the bottom with surprising energy and often displace large quantities of silt. This attracts other predatory fish which catch the startled bottom-dwellers. Young goatfish, surprisingly, develop in the open sea. They are silvery in colour and have a dark blue back. It is not clear how they protect themselves from predators in the open sea. They do not settle in the reefs until they have reached a length of several centimetres.

Food: Small fish and invertebrates.

Reproduction: Many species appear to spawn in pairs or groups at full or new moon. They prefer particular sites on open reefs. In this way the eggs are carried out by the tidal currents into the open sea and cannot be decimated by the many plankton-hunting reef-dwellers. The open sea is much more thinly populated, which increases the eggs' and larvae's chances of survival.

Two-barred goatfish, *Parupeneus bifasciatus*; up to 32 cm. Some species spend the day resting on the bottom. East Africa to Tahiti. Photo: Maldives.

Dash-and-dot goatfish, *Parupeneus barberinus*; up to 50 cm. The typical barbels can be seen in the picture. East Africa to Tahiti. Photo: Maldives.

Batfish – Platax

Genus *Platax*, Family Ephippidae, Order Perciformes. Size: up to 50 cm in length. There are four species in this genus. **Characteristics**: Batfish are strongly compressed laterally and extremely deep-bodied. With age, the depth changes in proportion to body length. Juveniles have very elongated dorsal, pelvic and anal fins, which become shorter with age. All species are silvery and have at least two horizontal stripes; one runs over the eye, the other through the base of the pectoral fins at the end of the gill cover, usually reaching the pectoral fins. The mouth is at the end and small. To identify the species: *Platax teira* has a strongly rounded dorsal fins and a convex head profile. The fins are markedly elongated in juveniles, the colouring is the same. *Platax pinnatus* is remarkable for its concave snout; the dorsal fin is relatively pointed. In juveniles the body is dark brown, bordered by a bright orange stripe which runs over the head and all the fins. This colouring is intended to imitate that of inedible nudibranchs and thus deter predators. The extremely elongated fins look out of proportion. As they grow, they lose their dark colouring and the orange stripes. *Platax orbicularis* resembles a crescent moon, as the fins curve backward until they are fully grown. Juveniles have a reddish-brown protective colouring which resembles a withered mangrove leaf. When in danger they flip over onto their sides and imitate a leaf bobbing up and down in the swell, as leaf fishes have long been observed to do.
Found: In tropical seas only. Usually in small groups, close to the reef, but also in the open water above the reef in large schools.
Way of life: Batfish are diurnal creatures which inhabit different territories. *Platax teira* is usually to be found in the open water above reefs. *Platax pinnatus* prefers the protection of the coral reefs and *Platax orbicularis* lives on sandy bottom if there is a reef in the vicinity to give shelter at night. Timidity also varies from one species to another. *Platax teira* is rarely afraid of humans and often keeps divers company.
Food: Jellyfish, small crustaceans, worms and invertebrates. Sometimes they will allow themselves to be fed with bread or hard-boiled eggs. It is however doubtful if this unusual diet benefits the fishes.
Reproduction: Only a little is known; one American species is known to spawn in groups of 10-20 individuals at a distance of some 40 m from the coast.

Pinnate Batfish, *Platax pinnatus*; up to 42 cm. Left: juveniles have markedly elongated dorsal and anal fins. Very young fish are dark brown with a bright orange border. Right: adult fish. Red Sea to New Caledonia. Photos: Thailand and Egypt.

Circular batfish, *Platax orbicularis*; up to 50 cm. they can be identified by their yellow pelvic fins. Red Sea to Tahiti. Photo: Thailand.

(bottom right)
Longfin batfish, *Platax teira*; up to 48 cm. The fish is at a cleaner station. Only juveniles have long fins. Red Sea to Australia. Photo: Thailand.

Butterfly fishes – Chaetodontidae

Family Chaetodontidae (previously a sub-family, with the angelfishes, of the Chaetodontidae). Order Perciformes. 113 species. Size: between 7 and 25 cm in length.

Characteristics: Body extremely laterally compressed and very deep; head relatively small. The large eyes are often set in a vertical dark stripe. The more or less pointed snout ends in a tiny mouth set with bristle-like teeth. The long-nosed butterfly fishes have pipe-like extended snouts. In most species the continuous dorsal fin is rounded off at the back. The first spines of the dorsal fins are linked with cut-out membrane. The rear of the dorsal fin and the anal fin are similar in size and shape in many species. The tail fin is convex in form, straight in exceptional cases. The pectoral fins are almost always colourless and transparent. The bodies of butterfly fishes are covered with scales. The splendid colours and patterns have earned these creatures the name of butterfly fishes. The brilliant colours in this family do not provide good camouflage, but that does not seem to endanger the continuation of the species. The conspicuous distinguishing features are important in a habitat as densely populated as a coral reef – both for reproduction and for territorial behaviour.

Some species have a spot on the back part of the body which is supposed to look like an eye. It is assumed that this eye spot together with the eye stripe which disguises the eye confuse attacking fishes. They are deceived as to the supposed direction of flight. This delay is usually enough for the butterfly fish to take flight among the corals, where the much bigger predators cannot follow.

Found: In tropical seas, generally near coral reefs. Juveniles prefer shallow water and like to stay among the protective branches of *Acropora* corals.

Way of life: Butterfly fishes, with their compressed bodies, are well adapted to life in the coral reef. They can manoeuvre skilfully through narrow cracks and the branches of corals. Their deep body build provides them with good lateral stability. They glide slowly and apparently effortlessly through the coral reefs, without making conspicuous movements. They almost always swim using only their pectoral fins, which are so transparent that they are hardly noticeable.

Saddled butterfly fish, *Chaetodon ephippium*; up to 23 cm. Lives singly or in groups in coral-rich reefs. Thailand to Tahiti. Photo: Thailand.

Klein's butterfly fish, *Chaetodon kleinii*; up to 13 cm. Found in loose groupings at the reef edge at depths of 2-25 m. East Africa to Hawaii. Photo: Maldives.

Latticed butterfly fish, *Chaetodon rafflesii*; up to 14 cm. Lives in areas with strong coral growth. Sri Lanka to Thailand. Photo: Thailand.

Lemon butterfly fish, *Chaetodon semilarvatus*; up to 23 cm. Lives in pairs or groups at depths of 2-30 m. Red Sea, Gulf of Aden. Photo: Egypt.

Sickle butterfly fish, *Chaetodon falcula*; up to 20 cm. Common in coral-rich reefs. Depth: 1-20m. East Africa to Indonesia. Photo: Thailand.

Meyer's butterfly fish, *Chaetodon meyeri*; up to 16 cm. Found in pairs in protected, coral-rich reefs. East Africa to Tahiti. Photo: Thailand.

Butterfly fish (continued)

Butterfly fish are diurnal and hide in the reef by night; they often change their colouring when doing so. Different butterfly fish have different specialised feeding habits. Some species living in couples depend on coral polyps and have territories in shallow water. Species that occur in shoals feed on zooplankton along the reef slopes at medium depths.

Although almost all fish are hermaphrodites, there are some species among the butterfly fishes where the sexes are separate and remain unchanged. Many species are monogamous and live together for long periods, if not for life.

Food: Plankton, coral polyps, crustaceans, worms, jellyfish, fish eggs or algae. There are many species in this family with specialised feeding habits.

Reproduction: Butterfly fishes swim to the surface at twilight to lay their eggs. The male butts the female in the belly. The eggs are laid in the water and drift away as plankton. The planktonic larvae hatch in only two days. The larval stage lasts for weeks or months. When it is dark the larvae settle in the protection of the reef.

Red Sea racoon butterfly fish, *Chaetodon fasciatus*; up to 18 cm. In the similar species Ch. lunula, the moon crescent butterfly fish, the white band is longer. Red Sea to Gulf of Aden. Photo: Egypt.

Red Sea chevron butterfly fish, *Chaetodon paucifasciatus*; up to 14 cm. Only found in the Red Sea. Photo: Egypt.

Long-nosed butterfly fish, *Forcipiger flavissimus*; up to 19 cm. The very similar species F. longirostris has dark scales on the breast and a longer mouth. Red Sea to eastern Pacific. Photo: Thailand.

Long-fin butterfly fish, *Heniochus diphreutes*; up to 22 cm. Always lives in schools in protected coral reefs. East Africa to the Society Islands. Photo: Thailand.

Masked butterfly fish, *Heniochus monoceros*; up to 21 cm. Found in pairs. Shown here with the cleaner fish *Labroides dimidiatus*. East Africa to Tahiti. Photo: Maldives.

Indian Ocean Bannerfish, *Heniochus pleurotaenia*; up to 18 cm. Found motionless in pairs or in groups at certain sites on the coral reef; easily confused with *H. varius*. Maldives to Java. Photo: Maldives.

Angelfishes – Pomacanthidae

Family Pomacanthidae (previously included as the sub-family Pomacanthinae among the Chaetodontidae, due to their close relationship with butterfly fishes). Order Perciformes. Some 80 species. Great variation in size: the smallest dwarf angelfish species is only 4.5 cm long, the largest can grow to 60 cm.

Characteristics: Angelfishes are distinguished from butterfly fishes by the typical, powerful backward-pointing spine on the gill plate, just in front of the gills. This is sometimes brightly coloured and in all species it begins at the bottom of the gill cover (see diagram). The mouth is less pointed, and with its thicker lips looks like a "pout".

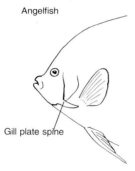

Angelfish

Gill plate spine

Angelfishes are often more strikingly coloured than butterfly fishes. The first spines of the dorsal fin are considerably shorter than those of butterfly fishes. Pectoral fins are colourless in only a few species. In angelfishes the shape of the tail fin varies from rounded to crescent-shaped. As they grow angelfishes undergo astonishing colour changes. Juveniles of the genus *Pomacanthus* are dark blue to black and have white, pale blue or blue-violet stripe patterns which differ from species to species. These patterns change so much that juveniles do not appear to have anything in common with adults. This is why the juveniles previously were given a different scientific name from the adults. Shortly before reaching sexual maturity, the juveniles change colour and patterning within a few weeks (see the series of pictures *P. imperator*). During this time both patterns can often be seen at once (see picture centre right). This change of colour is necessary as adult angelfishes have territorial behaviour patterns due to their feeding habits, and juveniles would not have any chance of survival set against the bigger fully-grown fish.

Emperor angelfish, adult (see also centre pictures). Photo: Egypt.

Emperor angelfish, *Pomacanthus imperator*, up to 35 cm. Above: Adult. Centre left: juvenile. Centre right: in-between stage. Red Sea to Tahiti. Photos: Egypt and Maldives.

Blue-ringed angelfish, *Pomacanthus annularis*; up to 35 cm. Shown with suckerfish. Prefers rocky reefs. Andaman Islands to west Pacific. Photo: Thailand.

Blue-faced angelfish, *Pomacanthus xanthometopon*; up to 40 cm. Usually single or in pairs along outer reefs. Maldives to Australia. Photo: Maldives.

Angelfishes (continued)

Juveniles and fish in the process of colour change are hard to find, as they are very timid and live in hiding. Males and females of the genus *Genicanthus* have different patterns and colours (see picture on the right).

Found: Mostly as inhabitants of tropical waters, both in coral and in rocky reefs. Only a few species occur in sub-tropical waters.

Way of life: Angelfishes are diurnal and generally live as individuals or in pairs. Like the butterfly fishes, they swim with their pectoral fins. Usually a male will have a harem of 2-5 females in a relatively large territory, but they will not be together all the time. The males can hold a territory of a few square metres (*Centropyge*) to 1000 square metres (*Pomacanthus*). This is defended vigorously against members of the same species. This territorial behaviour secures the food supply: the genus *Centropyge* subsists mainly on sea-bed algae, while *Genicanthus* feeds on zooplankton and yet other genera eat sponges, fish eggs or invertebrates. All angelfishes start life as females and can if necessary change to males. The process cannot be reversed. Young angelfishes have been observed acting as cleaner fishes and eating parasites off older fish. Angelfishes can make loud noises in the water. If disturbed, they issue a sound like "crack", intended to scare away the attacker. This threatening behaviour is probably also used at the borders of its territory against other members of the same species.

Food: Algae, sponges, small invertebrates. There are however species which do not despise fish eggs, zooplankton and bottom-dwelling invertebrates.

Reproduction: Egg laying usually takes place in pairs at sunset in open water, with the male swimming up and down and then butting the female in the belly. The planktonic larvae hatch within 24 hours and settle in the reefs after three to four weeks.

Zebra lyretail angelfish, *Genicanthus caudovittatus*; up to 25 cm. Male (female shown in the picture centre left). Lives in pairs along plankton-rich, steep reef slopes at depths of 15 to 45m. Red Sea. Photo: Egypt.

Zebra lyretail angelfish, female (see also above picture). Photo: Egypt.

Three-spot angelfish, *Apolemichthys trimaculatus*; up to 30 cm. In coral and rocky reefs, usually solitary. East Africa to Samoa. Photo: Maldives.

Regal angelfish, *Pygoplites diacanthus*; up to 26 cm. This attractive fish prefers calm, protected and coral-rich reefs. Red Sea to Tahiti. Photo: Maldives.

Moon angelfish, *Pomacanthus maculosus*; up to 50 cm. Red Sea to East Africa, Persian Gulf. Photo: Egypt.

Damselfishes and Anemone fishes – Pomacentridae

Family Pomacentridae, Order Perciformes. More than 200 species in four sub-families. Size: 4-20 cm in length.

Characteristics: Laterally compressed, usually deep-bodied with small blunt head, relatively high-set eyes; continuous dorsal fin and deeply forked tail. Colour varies greatly. The sub-family Pomacentrinae are black and white, often with horizontal stripes; but this colouring is also present in the sub-family Chrominae, in the genus *Dascyllus*. In the other sub-families, except for the anemone fishes, only the juveniles are generally brightly coloured. All species have clearly visible large scales.

Found: World-wide in tropical and sub-tropical seas. Most common in shallow water in coastal regions. They like to stay where there are plenty of hiding places, e.g. among corals of the genus *Acropora*.

Way of life: Pomacentrids are diurnal and behave territorially when fully grown. As juveniles they are found in large numbers snatching plankton above "their" coral tree. In danger and at night they hide among the coral branches. It is assumed that every fish has its own particular spot. This would mean that territorial behaviour is present in youth but a small living space is shared with others of the same species for protection purposes. The algae feeders among the pomacentrids are particularly aggressive. There are some species which act as "farmer" fish, looking after their own algae farms and plucking out varieties that do not taste good. Juveniles of the species *Dascyllus trimaculatus* look for protection in sea anemones just like anemone fishes.

Food; Mainly plankton, some also eat algae. There are also omnivores.

Reproduction: Pomacentrids often spawn in the morning, some species at the height of a particular phase of the moon. The stalked, elliptical eggs are fastened onto a firm base which has previously been thoroughly cleaned. The base is usually rock or coral rock; dead coral branches or a sandy bottom are also possible. The eggs can be of many different colours: red, pink, violet, brown, green, white or transparent. As a rule the eggs are watched over and cared for by the male. The eggs are provided with fresh water for oxygen by fin movements. Eggs which have died are carefully removed by the parent on guard. The planktonic larvae hatch at night, swim upwards and drift with the current towards the open sea.

Sergeant-major, *Abudefduf vaigiensis*; up to 17 cm. Often found in large shoals. All tropical seas. Photo: Thailand.

Philippine damsel, *Pomacentrus philippinus*; up to 9 cm. A widespread and common species. Thailand to Fiji. Photo: Thailand.

Black axil chromis, *Chromis atripectoralis*; up to 10 cm. The very similar *Ch. viridis* has no black spot on the inside of the pectoral fin base. Madagascar to Tahiti. Photo: Thailand.

Anemone fishes – Amphiprioninae

Sub-family Amphiprioninae, Family Pomacentridae, Order Perciformes. 27 species Size: 8-15 cm in length.

Characteristics: Body laterally compressed, relatively deep, blunt head and small mouth at the end. The fins are rounded, the body covered with scales. Many anemone fishes have very contrasting colours. Nearly all species are marked by one to three white vertical stripes. Anemone fishes can only survive in combination with certain large sea anemones. Juveniles change their patterns and their colouring as they develop.

Found: In tropical seas; Red Sea to Tahiti.

Way of life: Anemone fishes live in symbiosis with "their" anemones. They rarely move more than 2 m away from the anemone and when in danger can look for protection among the stinging tentacles. On the other hand, they also protect the anemone's tentacles against specialised predators which feed on them. The little anemone fishes are not afraid of quite large aggressors and will even attack divers who come too close to the anemones. The stinging poison cannot harm them, as they begin smearing the anemone's secretions onto their bodies in "childhood" and when they come in contact later, the anemone fishes are "immune". When the anemone touches the anemone fish it reacts as if the tentacles have touched each other. If a strange fish touches the tentacles, it is held and has no chance of escape. Each species of anemone fish prefers certain anemones. All juveniles begin as males and can change sex if required. A strict hierarchy operates within each anemone. The largest anemone fish is always a dominant female. The second largest is a male. All the others are smaller and fully grown fish prevent their growing further. This process has not yet been fully researched. It would be possible to speak of a "social repression". If the female dies, the highest ranking male becomes a female within a few weeks and the "second largest" male soon catches up in growth.

Food: Plankton and small crustaceans.

Reproduction: Anemone fishes glue their eggs to a firm base at the foot of the anemone: the eggs are guarded by the male. After a week the larvae hatch and drift away with the current as plankton. After two to three weeks the larvae settle in a reef and the juveniles begin to look for an anemone. Due to the short larval stage many species are endemic, i.e. they occur in a limited area.

Anemone Demoiselle, *Amphiprion ocellaris*; up to 10 cm. India to Indonesia. Photo: Philippines.

Saddle anemone fish, *Amphiprion ephippium*; up to 12 cm. Juveniles have a white neck band. Thailand to Java. Photo: Thailand.

Clark's anemone fish, *Amphiprion clarkii*; up to 12 cm. Persian Gulf to Fiji. Photo: Maldives.

Two-banded anemone fish, *Amphiprion bicinctus*; up to 10 cm. Red Sea and Gulf of Aden. Photo: Egypt.

Black-finned anemone fish, *Amphiprion nigripes*; up to 11 cm. Maldives, Sri Lanka and Lakkadives. Photo: Maldives.

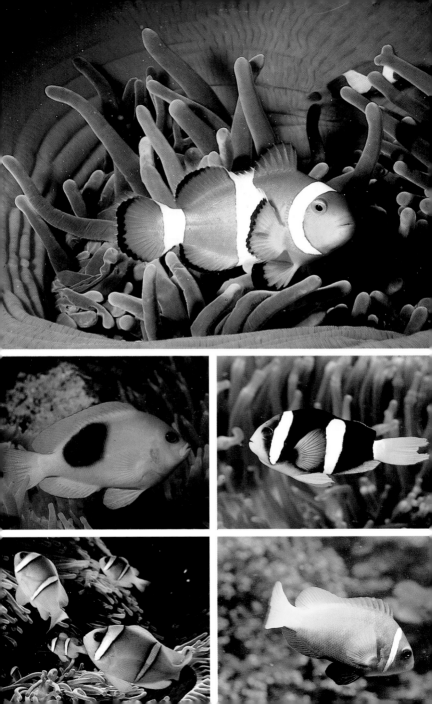

Wrasses – Labridae

Family Labridae, Order Perciformes. about 500 species according to rough estimates. They vary greatly in size, with the smallest only reaching 4 cm in length, the largest more the 2 m. Most species reach a size of up to 25 cm.

Characteristics: In body shape the wrasses vary as much as in size. The few large species have a deep laterally compressed body while the smaller ones tend to be small and elongated. All of them have a continuous dorsal fin and scales. Males and females differ in colour and pattern; male wrasses are usually brighter and more conspicuous. Juveniles undergo a series of colour changes during their development, which makes the classification of species much more difficult. Even today not all the juveniles have yet been definitively classified. In addition, many species change colour in the spawning season. It is even possible that two differing colorations will appear in individuals of the same species and sex (e.g. in *Epibulus insidiator*).

The shape of the body can also vary from one sex to the other. The many small species of wrasse generally have a small head and an apparently small mouth at the end. The mouth can often be everted (see diagram) or opened relatively wide. Wrasses have well developed dentition, including tearing and grinding teeth. The larger species are squat with big heads and thick lips. Older fish often develop a protuberant ridge on the forehead, e.g. the napoleon fish.

Napoleon fish, *Cheilinus undulatus*; up to 230 cm. Very tame in the Red Sea. Red Sea to Tahiti. Photo: Egypt.

Red-breasted wrasse, *Cheilinus fasciatus*; up to 30 cm. Red Sea to Samoa. Photo: Maldives

Wrasse: everted mouth

Wrasses and their nearest relatives, the parrotfishes, are easily recognised by the see-saw motion of their swimming. They swim with powerful strokes of the pectoral fins, which creates the slight up-and-down movement that is typical for this group.

Found: World-wide in tropical, sub-tropical and temperate seas.

Labridae (continued)

Way of life: Wrasses are diurnal and sleep at night. Many species dig themselves into the sand or hide among stones, others sleep in crevices or caves. Small wrasses are speedy and highly mobile swimmers and are constantly in motion, while the larger ones swim calmly along. Many cleaner fishes belong to the wrasse family (see also page 28, symbiosis). The smallest species of the genus *Labroides* can often be seen in tropical seas performing this task. They live within the bounds of a small territory which is visited by the most diverse range of fish species for "body care" purposes. Cleaner fishes do not just search the body surface but also the mouth and gills for parasites, on which they feed. Large fishes such as sharks and mantas are also cleaned by large species of wrasse.

Wrasses can change their sex. When the fish reach sexual maturity they enter the so-called starter phase, in which they are female and not so intensely coloured. Later, in the so-called final phase, they mature into splendidly coloured males. Juveniles often resemble the females. The confusing changes of colour which accompany the change of sex have led to many different ages having been classified previously as different species.

Food: Mainly fish and invertebrates. Some species feed on polyps or plankton, while cleaner fishes eat the parasites of other fish.

Reproduction: Wrasses generally spawn in open water above the reefs. A lengthy sequence of courtship behaviour can be observed, including various forms of "dance". Juveniles form large groups to spawn, older fish spawn in pairs. Some species build nests above which the male swims back and forth many times, fertilising the eggs. The eggs are also guarded by males.

Barred thicklip wrasse, *Hemigymnus fasciatus*; up to 50 cm. This colour variation however could indicate a previously unclassified species. Red Sea to Tahiti.

Sling-jaw, *Epibulus insidiator*; up to 35 cm. This species can evert its mouth (see diagram page 166). Red Sea to Hawaii, Tahiti. Photo: Egypt.

Bluestreak cleaner wrasse, *Labroides dimidiatus*; up to 10 cm. The best-known cleaner fish. There are often 2-5 fish at a cleaner station. Red Sea to Tahiti. Photo: Maldives.

Diana's hogfish, *Bodianus diana*; up to 25 cm. Juveniles look quite different; brown with white spots. Red Sea to Samoa. Photo: Maldives.

Wrasses – Labridae

1. **Two-tone wrasse**, *Thalassoma amblycephalum*; 14 cm. Juveniles and females have a black lateral stripe. Active, fast swimmers. Males have harem of several females. Inhabits lagoon reefs and outer reefs (depth of 20 m). Large gatherings of females often seen in shallow water. Feeds mainly on crustacean zooplankton. Maldives to Tahiti.

2. **Red-cheeked wrasse**, *Thalassoma genivittatum*; 20 cm. This rare species is only found in Mauritius and off the coast of Natal.

3. **Saddle wrasse**, *Thalassoma duperrey*; 25 cm. Females have similar colours but less intense. Very common in inner and outer reefs from the breaker zone to depths of 23 m. Sometimes observed as cleaner fishes. Endemic around Hawaii and Johnston Islands.

4. **Gold-bar wrasse**, *Thalassoma hebraicum*; 23 cm. Females are paler in colour. Males usually solitary, females in small groups. Inhabit sheltered inner reefs up (depth of 18 m). East Africa to Maldives.

5. **Sunset wrasse**, *Thalassoma lutescens*; 30 cm. Juveniles yellow, with black eye spots in the dorsal fin. Lives in clear lagoons and outer reefs to depths of 30 m. Also inhabits open sandy or stony areas and dense coral growth. Feeds on small benthic invertebrates. Sri Lanka to Panama.

6. **Crescent-tail wrasse**, *Thalassoma lunare*; 25 cm. A glowing bluish-green, with yellow-red tail. Prefers lagoons and coastal reefs (depths to 20 m), in the upper levels of overhangs and heads of coral. Feeds mainly on small bottom-dwelling invertebrates. Red Sea to Line Islands.

7. **Jansen's wrasse**, *Thalassoma janseni*; 20 cm. Black and yellow basic pattern does not vary much during growth. They live in sheltered lagoon reefs and exposed outer reefs with medium density of coral (depths to 15 m). Normally solitary. Maldives to Fiji.

8. **Six-bar wrasse**, *Thalassoma hardwickii*; 20 cm. Female less strongly coloured. This common species inhabits shallow lagoons and outer reefs (depths to 15 m) in clear water. Food: Crustacean plankton, small fish and foraminiferans. East Africa to Tuamotus.

9. **Surge wrasse**, *Thalassoma purpureum*; 43 cm. Female has brown stripes on a pale background. Prefers the breaker zones on the reef platform, reef edge and rocky costs. Feeds on fish and bottom-dwelling invertebrates. Red Sea to Easter Island.

10. **Ladder wrasse**, *Thalassoma trilobatum*; 29 cm. Female similar to *T. purpureum*, but with horizontal lines behind the eyes. Solitary dweller on the reef plateau, reef edge and rocky coasts. Prefers the breaker zone. Food as above species. East Africa to Pitcairn Islands.

11. **Five-stripe wrasse**, *Thalassoma quinquevittatum*; 16 cm. Similar is *T. klunzingeri* (found only in the Red Sea). Common in reef channels, lagoons and outer reefs (depths to 18 m). East Africa to Hawaii.

12. **Black-tail wrasse**, *Thalassoma ballieui*; 39 cm. Female paler in colour. Common in clear lagoons and outer reefs of rocky coasts with little coral growth. Found only around Hawaii and the Johnston-Islands.

Barracudas – Sphyraenidae

Family Sphyraenidae, Order Perciformes. About 20 species. Size: up to 1.80 m in length.

Characteristics: Elongated, almost cylindrical body covered with little scales. Large eyes, huge mouth studded with terrifying dagger-like teeth. The lower jaw protrudes noticeably. The well developed tail fin is slightly forked. The two dorsal fins are well separated. All species are silvery and may also have indistinct vertical stripes which can sometimes only be seen on the top half of the body. Juveniles usually have indistinct dark or sometimes yellowish horizontal stripes.

Found: World-wide in tropical and sub-tropical seas; often in bays and estuarine areas.

Way of life: Barracudas are very fast predators which are solitary as adults. There are diurnal and nocturnal species; the latter, and juveniles, spend they day inactive in large shoals. Barracudas camouflage themselves by placing themselves upright or at a slant among coral branches of about their size. Many divers report that they never see barracudas coming; "suddenly they're there." This is because, seen from the front, they show hardly any contrast and are well camouflaged.

Food: All kinds of fish.

Warning: in some areas barracudas are more feared than sharks. Their teeth make wounds which can be treated only with great difficulty. Victims can therefore, it is said, often bleed to death. Attacks are probably due to optical illusions in the water. Barracudas react to glittering objects such as fish which have been harpooned. Attacks are however very rare. It has been observed that fish make threatening bite movements before attacking. This could be territorial behaviour. Sharks are also known to make threatening gestures. If one is familiar with this behaviour and makes a determined exit from the barracuda's territory, an attack can probably be prevented. Barracudas are very curious and approach divers. In clear water there is no reason to be afraid of them. When eating barracuda, cases of ciguatera poisoning have been known in some case. In some parts of the Caribbean barracuda may not be sold for this reason. As a rule, no barracuda over the length of 1 m should be eaten.

Great barracuda, *Sphyraena barracuda*; up to 2 m. In some areas they are more feared than sharks. Great barracudas occasionally make threatening mock attacks. All tropical seas. Photo: Thailand.

Yellowtail barracuda, *Sphyraena jello*; up to 75 cm. Juveniles live in shoals. East Africa to Marquesas Islands. Photo: Thailand.

Blennies – Blenniidae

Family Blenniidae, Order Perciformes. Two Sub-families: Salariinae, comb-toothed blennies (see upper picture) and Blenniinae, sabre-toothed blennies (lower picture). About 300 species. Size: up to 15 cm.

Characteristics: Slender body with small head. Combtooth blennies have a blunt head with various attachments, a mouth at the end and usually have large, high-set eyes. The head of the sabretooth blennies, by contrast, is somewhat more pointed and their mouth is underneath. All have a long continuous dorsal fin and a relatively long anal fin. Combtooth blennies use their pectoral and pelvic fins to support themselves and hold their position. They have no swim bladder or scales. Their colour and patterns vary greatly; males and females also differ. Some are well adapted to the sea bed while others are conspicuously coloured or patterned.

Found: World-wide in tropical and temperate seas, in rocky coastal regions and coral reefs.

Way of life: The way of life of both sub-families of blennies is very different. The combtooth blennies are typical bottom dwellers which hold on to the bottom and only swim very short distances. They are mainly herbivores but do not disregard small creatures which hide among the algae. When in danger they withdraw backwards into their burrows. The sabre-toothed blennies are fast and enduring swimmers and are predators. The best known of them is the false cleaner fish *Aspidontus taeniatus* whose colouring mimics the cleaner wrasses of the genus *Labroides*. This sort of similar colouring often occurs in nature and is known as mimicry. The false cleaner fish not only resembles the genuine one in colouring, it also imitates its behaviour, in order to approach its victims unobserved. The trusting fish which believe they are going to be "cleaned" then have pieces of skin or scales torn off by the blenny's sharp teeth. Blennies live in one place and inhabit the burrows of tube worms and other creatures. Some also build their own burrows. Males are territorial.

Food: Combtooth blennies; mainly algae. Sabretooth blennies are predators.

Reproduction: Blennies lay eggs which stick to surfaces in cracks, caves, under stones or in shells. Eggs are generally guarded by the male but sometimes by both parents. Most species have a short larval stage.

Blackflap blenny, *Cirripectes auritus*; up to 7 cm. Male. Typical feature is the dark, yellow-bordered spot above the gill cover. East Africa to the Philippines. Photo: Thailand.

Blue-striped blenny, *Plagiotremus rhinorhynchus*; up to 11 cm. This is one colour variation: juveniles and female of this species imitate the blue-striped cleaner wrasse (see picture page 169). Red Sea to Marquesas Islands. Photo: Thailand.

Sleepers – Ptereleotrinae

Sub-family Ptereleotrinae, Family Microdesmidae, Sub-order Gobioidei, Order Perciformes. 150-250 species. Length: up to 12 cm.

Characteristics: Elongated slender body, small blunt head with large eyes and slightly protruding lower jaw. The mouth, pointing diagonally upwards, is larger in many males than in females. The first dorsal fin of the dartfish *Ptereleotris* has one or two greatly elongated spines bordered by a membrane and resembling the fins rays. The well developed second dorsal fin and the anal fin are about equal in size and shape and give the fish their very characteristic profile. Some species are conspicuously colourful while others are uniform in colour – generally bluish. During mating males may change their colour. The species *Ptereleotris zebra* has a pattern of vertical stripes in a delicate red-violet.

Found: Widespread in the Indian and Pacific Oceans but also in some places in the Caribbean; above sandy, stony or rocky beds.

Way of life: Some species spend their time resting on the bottom of shallow reefs: hence their name. Others are active swimmers which spend their time almost exclusively in the open water. At night or when in danger they seek refuge in their burrow. By day they hover about 1 m above the sea bed, not far from their hiding places, and snatch at passing plankton. They usually live in pairs, sometimes singly, at depths of 2-50 m. Juveniles can also be seen in groups.

Species living above sand will dig themselves into the sand when in danger. Sometimes they will also build burrows. Dartfish can be identified by their jerky style of swimming. They hover motionless in one spot, then suddenly see and snatch some plankton and then remain still again until they see they next piece of food. They even stay in the same spot in a current, without noticeable swimming movements. When enemies approach the sleepers stay close to their burrows. Should the enemy get too close, the fish moves head over heels into its refuge, so quickly that the eye is hardly able to follow it.

Food: Zooplankton.

Reproduction: Little is known. The species which have been studied show strong territorial behaviour, with the male not letting the female out of his sight so that there can be no possibility of "adultery".

Fire dartfish, *Nemateleotris magnifica*; up to 6 cm. Lives in burrows and when feeding hovers half a metre above the ground. East Africa to Tahiti. Photo: Thailand.

Decorated dartfish, *Nemateleotris decora*; up to 7 cm. Lives at depths of 25 to 70 m. Mauritius to Samoa. Photo: Thailand.

Gobies – Gobiidae

Family Gobiidae, Sub-order Gobioidei, Order Perciformes. About 1600 species. Size: generally 2-15 cm; only very few grow longer than 30 cm.

Characteristics: Elongated or compacted, slightly conical body, blunt head with relatively large mouth. An unmistakable feature is the fused pelvic fins which form a single cup-shaped sucker fin with which the fish can attach itself to the bottom. Gobies usually have two dorsal fins and a rounded tail fin. They support themselves on the bottom with their large pelvic fins. Colour and pattern vary strongly, but most species are well adapted to the sea bed. Males are usually conspicuously coloured.

Found: In coastal waters of tropical and temperate seas, in water that is not too deep; on sandy or muddy bottoms or in coral-grown areas.

Way of life: Gobies are predators that "ambush" prey, lying in wait on the bottom. They are not strong swimmers. They can, however, swim relatively fast for short distances. In this large family some interesting specialists with different ways of life are to be found:

Plankton-eaters have a swim bladder and hunt in water near the surface.

Another group, sentinel gobies, live in symbiosis with snapper shrimps (see picture page 77). The shrimps live in burrows in sandy or stony bottom which they have dug themselves. They can see little or may even be blind. The gobies, therefore, take on the function of sentinels and lie in front of the burrow entrance. The snapper shrimp is continuously busy removing grains of sand which have slid back into its burrow. Sometimes it also drags a larger stone out with its pincers or it uses the stone to push the sand out in front of it. As soon as it reaches the exit its long feelers touch the goby, which uses certain movements to signal to the shrimp that "the coast is clear". If danger threatens first the shrimp and then the goby flee into the protecting cave. The community always consists of two shrimps and one or two gobies. Without the sentinel goby the snapper shrimp could never leave the burrow, as it would immediately fall prey to a predatory fish.

The smallest species of goby, *Pandaka pygmaea*, only grows up to 1 cm in length and is therefore the smallest of vertebrates.

Food: Small fish, invertebrates and plankton.

Reproduction: During mating, the males of many species build nests in caves or under stones and care for the young.

Blue-streak goby, *Valenciannea strigatus*; up to 17 cm. Lives in burrows in sand or in coral rubble, usually in pairs. East Africa to Tahiti. Photo: Thailand.

Gorgeous dartfish, *Amblyeleotris wheeleri*; up to 7.5 cm. Lives in symbiosis with snapper shrimps (see also picture page 77 bottom right). East Africa to Marshal Islands. Photo: Thailand.

Surgeonfishes – Acanthuridae

Family Acanthuridae, Order Perciformes. About 50 species. Size: 13-55 cm (unicorn fishes are bigger).

Characteristics: Body relatively deep and laterally compressed; head and mouth generally small, forehead markedly rounded. A typical feature is the sharp continuation of the bone on either side of the tail base. These "scalpels" lie in grooves and only emerge during aggression. They are brightly coloured in some species. Surgeonfishes have long, continuous, large dorsal and anal fins which often follow their body shapes. The tail fins are often crescent-shaped. All species of the genus *Zebrasoma* have very high dorsal and anal fins: they were previously known as sailfish. Many surgeonfishes are conspicuous in colour and pattern, others can change colour during mating or at times of aggression. The colour depends on the mood of the fish. Some species change their colour so much during development that juveniles were previously considered a genus of their own (*Acronurus*). All species have small rough scales.

Found: World-wide in tropical seas, in relatively shallow water. Most common in the Indo-Pacific.

Way of life: Surgeonfishes are diurnal and sleep at night, hidden in the reef. They are slow swimmers as they move forwards using only their pectoral fins. Due to this style of swimming they have an up-and-down motion which gives them their typical rhythm. They are mainly herbivorous, spending the whole day scraping algae off coral rock. Some are solitary, others in small or larger groups which can consist of several hundred fishes. Their sharp scalpels are only used for defence and for rivalry battles.

Food: Predominantly algae, but also small bottom-dwelling creatures such as crustaceans, molluscs and worms, as well as detritus (floating and sinking matter).

Reproduction: During certain phases of the moon in the winter and early spring they spawn in pairs or groups at dusk. The eggs drift out to sea with the current. The larval stage is very long; they do not settle until they are relatively large.

Warning: The scalpels can cause deep, painful wounds, but they only serve for defence. If you do not try to catch the fishes they will not harm you. The same is also true of unicorn fishes.

Palette surgeon, *Paracanthus hepatus*; up to 25 cm. Juveniles prefer the Acropora corals. East Africa to Samoa. Photo: Thailand.

White-breasted surgeon, *Acanthurus leucosternum*; up to 30 cm. Often appear in great shoals above the reef plateau. Indian Ocean.

Goldring surgeonfish, *Ctenochaetus strigosus*; up to 16 cm. Juvenile, which will turn beige, then brown. East Africa to Hawaii. Photos: Thailand

Blue-banded surgeonfish, *Acanthurus lineatus*; up to 30 cm. The similar Red Sea Surgeonfish *A. sohal* lives in the Red Sea. East Africa to Hawaii. Photo: Thailand

Unicorn fishes – Nasinae

Sub-family Nasinae, Family Acanthuridae, Order Perciformes. About 15 species. Size: up to 1 m long.

Characteristics: Unicorn fishes are bigger than their nearest relatives, the surgeonfishes. A few members of this sub-family have a long forward-pointing peak of bone (the "horn") on the forehead between the eyes. This horn will grow as they develop. In the species *Naso unicornis* only the male has a horn. Other species have a round swelling on the forehead. Unicorn fishes have two "scalpels" (see diagram) on each side, but these cannot be folded back. They do not start to grow until the fishes reach sexual maturity. Many species are inconspicuous in colour but can change colour depending on their mood.

Found: From the Red Sea to the eastern Pacific.

Way of life: Unicorn fishes are diurnal; at night they hide and rest. They are found singly or in shoals of varying sizes along the reef slopes.

Food: Zooplankton, some brown algae.

Reproduction: During the mating season the male displays conspicuous colours and patterns. The larval stage is very long.

Moorish Idols – Zanclidae

Family Zanclidae, Order Perciformes. Only one species. Size: up to 18 cm in length.

Characteristics: Similar to the bannerfish of the butterfly fish family (see page 156) in body build. Moorish idols can be clearly distinguished from these by their black tails. They are closely related to the surgeonfishes, but have no scalpels. The eyes are set high. The third spine of the dorsal fin is extremely elongated by a whip-like filament which protrudes about 10 cm past the tail. The adult fish grow small, curved horns between the eyes.

Found: From the Red Sea to Tahiti and Hawaii.

Way of life: Moorish idols are pectoral fin swimmers and therefore not very fast. They are diurnal and live in pairs or groups.

Food: Omnivorous, but mostly sponges.

Reproduction: They spawn at night or in the dawn twilight. The larvae hatch after only 30 hours. The pelagic larval stage lasts for a relatively long time. The young settle in the reefs when they have reached 6-7 cm.

Humpback unicorn fish, *Naso brachycentron*: up to 60 cm. Solitary fish which prefers sheltered reefs. East Africa to western Pacific. Photo: Maldives.

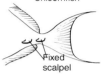
Orange spine Unicornfish

Fixed scalpel

Orange-spine unicorn fish, *Naso lituratus*; up to 40 cm. Night photograph. In the Pacific this species has a white dorsal fin. Red Sea to eastern Pacific. Photo: Maldives.

Moorish idol, *Zanclus cornutus*; 16 cm. Usually seen in small groups. East Africa to eastern Pacific. Photo: Maldives.

Rabbitfishes – Siganidae

Family Siganidae, Order Perciformes. 27 species. Up to 40 cm in length. They are related to the surgeonfishes.

Characteristics: Body laterally compressed, deep and longish. Small head with large eyes. The "bunny mouth" with the thickened lips is supposed to have given these fish their name. Their continuous dorsal fin is studded with 13 spines embedded in poisonous tissue. The first spine is small and points forward. The pelvic and anal fins are also equipped with poisonous spines. Rabbitfishes vary in colour. Small patterns of stripes and lines are distributed over the whole body and look like ornaments. *Siganus vulpinus* and *magnifica* have highly contrasting broad stripes on the head which give then a badger-like appearance. At night rabbitfishes take on a diffuse colouring. They can also change their colour quickly in stress situations. The scales are small and elongated. In some areas these are highly regarded as food fishes.

Found: In the Red Sea and Indian Ocean to the Pacific. (*Siganus rivulatus* has entered the Mediterranean through the Suez Canal.) Generally found in coral reefs at not too great a depth.

Way of life: Rabbitfishes are diurnal and spend the whole day grazing on the growth on firm sea bed, rocks or reefs. They eat a great deal and live in pairs or groups. At dusk they take on a diffuse colouring with hardly any contrasts and lie motionless on the bottom. This means they are relatively safe from predators, which usually hunt at dusk.

Food: Algae, seaweed, tunicates, sponges.

Reproduction: Rabbitfishes spawn in pairs or groups during certain phases of the moon in spring, usually at night or in the early morning. At low tide the eggs drift out to sea.

Warning: Nearly all fins are equipped with poisonous spines which can cause very painful wounds. Injuries are not dangerous, however. These weapons are only used for defence; these harmless fish are generally timid. Rabbitfishes are popular food fishes in all tropical countries and on sale in all fish markets. In some areas cases of ciguatera poisoning have occasionally been reported. In these areas people avoid eating rabbit fish.

Magnificent rabbitfish, *Siganus magnificus*; up to 25 cm. Lives in pairs in shallow reefs. At present known only in the Andaman Islands. Photo: Thailand.

Coral rabbitfish, *Siganus corallinus*; up to 28 cm. Grazes on carpets of algae in shallow reefs. Usually found in pairs. Seychelles to New Guinea. Photo: Thailand.

Tunas – Scombridae

Family Scombridae, Sub-order Scombroidei, Order Perciformes. 48 species. Size: 35-230 cm.

Characteristics: Spindle-shaped body with thin tail base and rounded cross-section; the head has a medium-sized mouth at the end and large eyes; the dorsal fins are clearly separated. The second dorsal fin and the anal fin are similar in shape and size. Behind these there is a row of small fins above and below reaching to the base of the tail, usually five to ten pairs (see diagram). The tail base is strengthened by powerful lateral ridges. Tuna are silvery in colour, darker on the back.

Found: World-wide in tropical and temperate seas; some also move into colder seas.

Way of life: Tunas are among the fastest swimmers and cover great distances. They can maintain their body temperature several degrees above that of the surrounding water. This "warm-bloodedness" means that they have great endurance. They are active by day and night and never sleep. They are found singly and also in large shoals. In the last few years the numbers have been greatly reduced by 50 km long drift nets.

Food: Squid, crustaceans and fish.

Tuna

Weevers – Pinguipedidae

Family Pinguipedidae, Order Perciformes. Few species. Size: about 15 cm.

Characteristics: Almost cylindrical, elongated body and small, conical head with mouth at end. They have noticeably protuberant eyes. The basic colours are beige or sand-coloured. A pattern of blotches covers the whole body from head to tail. The number of blotches is important for identification. Male and female fish have different patterns.

Found: In tropical seas; on sandy and stony sea beds.

Way of life: Weevers are predators that wait in ambush, lying motionless on the bottom. The head is always slightly raised from the bottom, as they rest on their pelvic fins. Males are territorial and have a small harem. All juveniles are female and can later change into males.

Food: Small fish and invertebrates.

Reproduction: The eggs are laid in open water before sunset and drift out into the open sea. The larval stage lasts one to two months.

Triggerfishes – Balistidae

Family Balistidae, Order Tetraodontiformes. Many species. Length: 17-70 cm.

Characteristics: Body laterally compressed and deep: the side view is rhombus-shaped. The very large head runs to a conical point; the eyes are set well back and very high; the small mouth at the end has powerful teeth which can in part be seen. The first dorsal fin can be retracted into a groove and is not always visible. It consists of three powerful spines which become smaller further back and have a mechanical locking system (see below). The second dorsal fin and anal fin are fringing fins and are alike in shape and size. Triggerfishes are often conspicuously coloured and patterned. They have very large, strong scales which form a kind of armour. The juveniles of some species differ greatly in colouring from the adults.

Found: In tropical and temperate seas; in shallow, rocky coastal waters, coral reefs and seaweed.

Way of life: Triggerfishes swim by moving the fringing fins formed by the second dorsal and anal fins in a wave-like fashion. Not only can they swim horizontally, but also diagonally or even in a vertical position. If the swimming speed is not fast enough using the fringing fins, the tail fin is used to increase speed. This happens when the fish are fleeing or if another fish (or a diver) comes too close to the eggs without heeding the warning signs.

The breeding behaviour of triggerfishes is familiar to many divers. The male takes up position head-downwards, blowing oxygen-rich water onto the eggs. At the same time it observes its surroundings with rolling eyes. Anyone who is not familiar with this behaviour and approaches the nest will be clearly warned by a mock attack; the triggerfish swims up to the supposed attacker at great speed and turns away shortly before colliding. Anyone who does not immediately leave its territory at that point is then rammed at full speed and bitten. Divers should never swim upwards when leaving the fish's territory: that seems to infuriate it. The nest is often guarded by both parents.

Outside the breeding season triggerfishes are solitary, peaceful and rather timid. They are diurnal and sleep at night in caves or holes, wedging themselves with their first dorsal fin spine. The second spine is wedged against the first and lies in a small depression. It is now impossible to fold back the first spine (see diagram).

Titan triggerfish, *Balistoides virides-cens*; up to 70 cm. Become aggressive in the breeding season. Red Sea to Tahiti. Photo: Egypt.

Dorsal fin spines

Clown triggerfish, *Balistoides conspicilium*; up to 30 cm. When in danger, triggerfishes can wedge themselves (see picture). East Africa to Samoa. Photo: Maldives.

Orange-striped triggerfish, *Balistapus undulatus*; up to 28 cm. Photo: Maldives

Triggerfishes (continued)

A predator discovering the triggerfish in its cave cannot pull it out. In this wedged position the triggerfish can then defend itself with its teeth. It is so strong it can effortlessly bite through hard-shelled mussels.

Triggerfishes have developed a special technique for attacking sea urchins, which Hans W. Fricke observed and photographed. It bites off the formidable long spines of the diadem sea urchin, until it finds a strong spine with which it can lift the sea urchin. It swims up about one metre with the sea urchin and lets it drop. While the sea urchin slowly sinks to the bottom, the triggerfish swims underneath and bites into the urchin's poorly protected underside. Another hunting technique is to bowl over the sea urchin with a powerful jet of water, which the triggerfish can create with its mouth. This jet of water is also used to expose prey when looking for food in the sand, and also when making a nest.

Triggerfishes build very large nest hollows. This means that there is an open space around the eggs which leaves a good clear view. It makes it less easy for a predator to get close to the eggs without being seen. Once a triggerfish has chosen a site for its nest it starts blowing the sand outwards with a powerful jet of water. Any pieces of broken coral or stones that it finds under the sand are taken into its mouth to the edge of the hollow and dropped. With a great deal of labour, it creates a large hollow which can reach a diameter of one to two metres with large species.

Food: Fish, molluscs, crustaceans, corals, tunicates, sea urchins and other echinoderms. Some species will also take algae or larger zooplankton.

Reproduction: They build nests and lay adhesive eggs just before twilight. These are watched over by the male or by both parents. The larvae, when hatched, drift upwards and out to the open sea on the current.

Warning: The larger species of triggerfishes are very aggressive during the breeding season and will even attack divers that approach their nests. It is advisable to know the warning signs of these fish and to take note of them (see also "Way of life"). As soon as you recognise this situation, immediately swim on your back away from the nest. Never swim upwards! Bites in the fins or the backs of the knees are not uncommon!

Blue triggerfish,
Pseudobalistes fuscus; up to 48 cm. Juveniles have broad blue stripes on a yellow ground. Red Sea to the Society Islands. Photo: Maldives.

Picassofish, ➤
Rhinecanthus aculeatus; up to 25 cm. Found in shallow water above sandy and stony bottom. East Africa to Hawaii, Tahiti. Photo: aquarium.

Redtooth ➤ ➤
triggerfish, *Odonus niger*; up to 33 cm. Often found in large numbers on reef slopes; when in danger, withdraws into crevices or holes, with the forked tail remaining visible. Red Sea to Tahiti. Photo: Maldives.

Yellow margin triggerfish,
Pseudobalistes flavimarginatus; up to 60 cm. Makes large nesting hollows (up to 2 m across) in the sand. Red Sea to Tahiti. Photo: Egypt.

Filefishes – Monacanthidae

Family Monacanthidae, Order Tetraodontiformes. About 20 species. Length: 6-75 cm.

Characteristics: Body strongly compressed laterally, usually relatively deep. Body shape, eyes and fins are similar to those of the triggerfish; generally smaller. Both families are very closely related and mainly differentiated by the first dorsal fin; filefish have a characteristic spine. The spine is long and thin – sometimes set with little prickles – and set far forwards, right above the eyes in some species. A second, very small spine is embedded under the skin and not visible; it also locks into position as in the triggerfishes. The small mouth, set at the end of the head, is equipped with powerful teeth. The tail fin is very large but is almost always folded flat. The tail base is laterally compressed but relatively deep. Small species vary greatly in form and can look like sharpnose puffers, but the body is flatter. The clearest features for identification are the considerably longer second dorsal and anal fins. These fish have very rough scales which give them the name of filefish. Most species are inconspicuous in colour and can adjust themselves for camouflage purposes, for example to the colour of their background. One small species, *Oxymonacanthus longirostris*, is conspicuous in colour. The patterns different for almost all species.

Found: In tropical temperate seas; in coral reefs, seaweed and algae zones of shallow coastlines.

Way of life: Filefishes are slow swimmers which propel themselves by moving the second dorsal and anal fins in a wave-like manner. They live singly, in pairs or in groups. They prefer shallow reefs with clear water and plenty of coral, especially soft corals or the stone corals *Acropora* or *Pocillopora*. They live a secretive life in these places. They move skilfully among the branches of the coral and can swim backwards as well as forwards. One filefish, the black saddle mimic *Paraluteres prionurus*, shows interesting mimicry – it is almost identical in colour to the poisonous sharpnose puffer *Canthigaster coronata*.

Food: Coral polyps, small crustaceans, fish.

Reproduction: They lay adhesive eggs which are guarded by the male or the female.

Filefish are not food fishes – the flesh is inedibly bitter, though not poisonous.

Black saddle mimic, *Paraluteres prionurus*; up to 9 cm. The only species which can mimic two differently coloured sharpnose puffers. In the Andaman waters it loses the typical saddle blotches and is brown with pale spots. Photo: Thailand

Longnose filefish, *Oxymonacanthus longirostris*; up to 12 cm. Smallest filefish; lives among *Acropora* corals. East Africa to Samoa. Photo: Thailand.

Scribbled filefish, *Aluterus scriptus*; up to 90 cm. Largest species in this family. All tropical seas. Photo: Egypt.

Trunkfishes – Ostraciidae

Familyn Ostraciidae, Order Tetraodontiformes. 37 species. Length: 13-13 cm.

Characteristics: Body and head are completely surrounded by hard, angular, "trunk-like" armour. The armour is made up of hexagonal bony plates which lie fused just below the scaleless skin. Most trunkfishes have a rectangular cross-section, some are triangular. The armour has gaps only for mouth, eyes, gills, anal vent and fins. All fins are rounded and, apart from the tail fins, are relatively small. Trunkfishes have only one dorsal fin and no pelvic fins. The genus *Rhynchostracion* has a "nose". Trunkfishes have a very small mouth and thick lips. Colour varies and can be different in juveniles and adults; in some species males and females are also differently coloured.

Found: World-wide in tropical seas.

Way of life: Trunkfishes are slow but very manoeuvrable, well adapted to life in the reef. With their small fins they can turn on the spot like a helicopter. They propel themselves by moving the second dorsal and anal fins in wave-like fashion in the opposite direction to one another. This very individual method of swimming is known as "gondolier style". The pectoral fins make propeller-like movements to balance out the swaying motion thus created. When in flight, the tail fin is often put to use, which can boost the fish to amazing speeds. Because of their armour, trunkfishes have no gill covers. The circulation of water necessary for the oxygen supply is achieved by raising and lowering the floor of the mouth cavity. When searching for food, these diurnal fishes place themselves in a head downwards position and with their mouths create a jet of water which churns up the sand, exposing hidden food. Males are territorial and keep a harem of three to four females.

Food: Small invertebrates and algae.

Reproduction: They look for high sites on the reef to mate. From these sites they swim upwards into the open water and leave the eggs to the current.

Warning: When stressed the fish exude a highly toxic secretion when can be fatal to other fishes. Especially where space is limited, e.g. in an aquarium, this can lead to the death of all the inmates, including the toxin-producing trunkfish.

Cube trunkfish, adult (see also central picture). Photo: Maldives.

Cube trunkfish, *Ostracion cubicus*; up to 45 cm. Juvenile with typical colouring (for adult, see picture above). Red Sea to Hawaii, Tahiti. Photo: Thailand.

Spotted trunkfish, *Ostracion meleagris*; up to 19 cm. The female is dark brown with white spots. East Africa to eastern Pacific. Photo: Thailand.

Pufferfishes – Tetraodontidae

Family Tetraodontidae, Order Tetraodontiformes. About 115 species. Size: 6-90 cm.

Characteristics: Body rounded and plump; the head is reminiscent of a seal's. The mouth, set at the end, is small and equipped with extremely strong teeth that can easily chew up hard-shelled creatures. The scaleless body is elastic and extremely stretchable. The fish have no gill covers but have gill openings in front of the pectoral fins. The dorsal and anal fins are set far back and are alike in shape and size; these fish are "gondolier-style" swimmers (see page 198). All fins except for the tail fin are relatively small; there are no pelvic fins. Nearly all pufferfishes are inconspicuous in colour. Species of the sharpnose puffer genus (*Canthigaster*) have, as the name says, a relatively pointed head and are only a few centimetres long.

Found: In all tropical and sub-tropical seas; in coral and rocky reefs, also where seaweed grows.

Way of life: Pufferfishes are slow but manoeuvrable and are well adapted to life in the reef. Their ability to expand their body to several times its size in a short space of time is intended to frighten away aggressors. They suck huge quantities of water into a separate chamber near the stomach. Unfortunately many divers catch pufferfishes just for fun, because the creature looks so "photogenic" when blown up. They overlook the fact that the fish feels itself to be in deadly danger and can after a while die of this unnecessary stress. Things are even worse when this happens at the surface, as the pufferfish may also swallow air and then cannot dive again. If it survives the stress, it will drift away and has hardly any chance of relocating a protective reef. Usually they stay near the bottom and try to camouflage themselves.

Food: Crustaceans, fish.

Reproduction: They lay adhesive eggs; otherwise little is known.

Warning: Their powerful teeth can remove a finger. In Japan pufferfishes are considered a delicacy and are prepared by "fugu chefs" who have undergone a long period of training. The intestines, sexual organs and skin of these fishes contains one of the most powerful poisons (tetrodontoxin) produced in nature. If the fish is wrongly prepared most cases of poisoning lead to death in a short space of time. The degree of toxicity depends on the species, the area where caught and the season.

Black-saddled toby, *Canthigaster valentinii*; up to 9 cm. Imitated by the black saddle mimic (see page 197, top picture). Red Sea to Tahiti. Photo: Egypt.

Black spotted puffer, *Arothron nigropunctatus*; up to 33 cm. There are several colour variations;: yellow, brown and bluish-grey. East Africa to Hawaii. Photo: Thailand.

Star puffer, *Arothron stellatus*; up to 1 m. This species is found in two colour variations. Red Sea to East Pacific. Photo: Thailand.

Porcupine fishes – Diodontidae

Family Diodontidae, Order Tetraodontiformes. 19 species. Size: up to 85 cm in length.

Characteristics: The body appears triangular, in profile as well as seen from above. The slightly flattened blunt head with the small mouth at the end is remarkable for its extremely large eyes. Head and body are covered with backward-pointing spines laid close to the body which are only raised when the porcupine fish blows itself up. The fins are almost identical with those of pufferfishes. Dentition consists of two parts: there is a continuous dental plate in each jaw, capable of a powerful bite. Porcupine fishes are generally beige to brownish in colour. A few have typical dark eye spots.

Found: World-wide in tropical coral reefs, also at greater depths.

Way of life: Porcupine fishes – like their nearest relatives, the pufferfishes - are slow swimmers and just as skilled at manoeuvring their clumsy-looking bodies among the tangle of coral branches of a reef. These slow "gondolier style" swimmers can speed up their swimming considerably by using their tail fins. They are diurnal and hide at night in the reef. When in danger, they pump themselves up with water like pufferfishes and can considerably increase their volume: this makes their relatively long spines radiate outwards. In this way the eye spots become a means of terrifying aggressors. If a large predator should happen to eat the porcupine fish, it will stick in the predator's mouth. The porcupine fish, blowing itself up still further, will wedge itself into the mouth of its attacker so that the latter will never be able to get it out. Quite large sharks are supposed to have suffocated in this manner. These timid and harmless fishes always stay close to the bottom and hide as soon as an enemy appears. Often all that can be seen is the large head with its huge eyes in a cave; inexperienced divers are startled by the "monster", believing that they are facing a large or dangerous creature. Some species bury themselves in the sand up to the eyes.

Food: They prefer hard-shelled invertebrates but also eat crustaceans and sea urchins.

Reproduction: At dusk the females rise to the surface with one or more males and spawn into the open water.

Short-spine porcupine fish, *Diodon liturosus*; up to 65 cm. Often in hiding during the day. East Africa to Society Islands. Photo: Maldives.

Spotted porcupine fish, *Diodon hystrix*; up to 90 cm. All tropical seas. Photo: Thailand.

Warning: Porcupine fishes can bite through a finger with their powerful teeth.

Marine turtles – Cheloniidae

Family Cheloniidae, Order Testudines, Class Reptilia. About 12 species, up to 1.40 m in length.

Characteristics: They differ from land turtles (tortoises) mainly in their feet, which have developed into paddle or fin shapes.

Found: World-wide, except in cold seas.

Way of life: Marine turtles were formerly land reptiles. They are air breathers and as such must travel regularly to the surface to breathe. They can swim very fast with their front limbs; the rear limbs are used solely for steering.

Food: Molluscs, crustaceans, fish and algae.

Reproduction: In the mating season they travel huge distances, back to the regions in which they were born. They meet to mate just off the shore. Mating takes place in the water. The females have to come to land to lay their eggs. They use their hind legs to dig deep pits in the sun-warmed part of the beach, and lay 80-200 eggs in them. Afterwards the exhausted creatures shovel the sand back over the clutch and smooth it down with the belly of their shells. Then they return to the sea. They leave a distinctive trail behind them. Local people often raid the clutch, so that numbers are decreasing rapidly.

Hawksbill turtle, *Eretmochelys imbricata*; up to 90 cm. Can be identified by the jagged edge of its shells. Indo-Pacific. Photo: Maldives.

Green turtle, *Chelonia mydas*; up to 1.4 m. Feeds mainly on algae. Unfortunately the flesh is still used commercially. Indo-Pacific. Photo: Maldives.

Sea snakes – Hydrophiidae

Family Hydrophiidae, Class Reptilia. About 50 species. Size: 1-3 m in length

Characteristics: The hind part of the body is laterally compressed and ends in a paddle-shaped tail. The lung extends to the tail and, together with a slow metabolic rate, allows the snakes to stay under water for several hours.

Found: In the Indian Ocean and Pacific, also in brackish water.

Way of life: Sea snakes can dive down to 200 m and spend long periods searching for food. They are usually not aggressive towards humans.

Food: Small fish.

Reproduction: The young are generally born live in the water: some species lay eggs out of the water.

Banded yellow-lip sea snake, *Laticauda colubrina*; up to 1.5 m. Extremely poisonous, but not normally aggressive. Lays 5-6 eggs on sandy beaches. Indo-Pacific. Photo: Thailand.

Warning: All sea snakes are more poisonous than a king cobra, but venom is not injected with every bite. About 25% of bites are fatal.

Underwater photography

If you have the sort of ambition that overcomes obstacles, you will find plenty of scope for your activities in underwater photography. I would like to give some advice – in the form of tips gained from practical experience – on how to avoid the biggest and most frequent problems.

Before leaping into the waves with your camera, you should become familiar with basic facts about photography and the laws of optics, and you should if possible have appropriate practical experience.

If you can handle living creatures and are good with them, you will take home better pictures than someone who is nervous of them.

However, the most important condition to meet is diving skill. A beginner who has just completed his or her first diving course will have quite a lot of theoretical knowledge, but the movements do not become automatic until after some practice. Through practical experience you will learn how to recognise upward or downward flow or currents in good time and react quickly. As soon as you start looking through the viewfinder of a camera, you will register such things relatively slowly.

Photographing moving objects requires great concentration. This will distract you even more from everything which is going on around you. You could sink to the bottom and break off corals which may have taken years to grow. If you want to capture the beauty of nature in a photograph, you should also try to destroy as little as possible. It is therefore better not to start underwater photography until you have enough diving experience.

If you are considering buying an underwater camera, you should acquire some basic knowledge, in order not to regret the purchase later. The relatively expensive special equipment and the more difficult conditions in this medium mean that the costs per successful picture rise enormously. A photograph that would be problem-free out of the water can often only be taken underwater with difficulty and will frequently have to be rejected; at any rate after critical consideration.

Every person has different expectations. A perfectionist will be forever trying to improve on existing good photographs, while others might be happy with a few souvenir snaps. If the latter is the case, is it worth buying expensive special equipment? Many diving schools in holiday resorts sell very good underwater slides or postcards, which as souvenirs have almost the same value

In the end your choice of camera system will depend to a certain degree on the results you wish to achieve

Underwater cameras

There are always new developments in underwater cameras, accessories, and underwater housings to fit land cameras. Previously you could choose between a waterproof underwater camera with a viewfinder (Nikonos or

Sea and Sea) or a single lens reflex camera (SLR) which must be fitted with an waterproof housing. The latest development is a waterproof single-lens reflex camera, the Nikon RS.

By far the most expensive system is the Nikon RS with underwater flash and two lenses. An SLR with similar lens, flash and suitable housing will cost a third of the price of the RS, and the lowest-priced (Sea and Sea) only a fraction of the RS camera. Neverthe-less, when making a purchase it is not just the price tag but the advantages and disadvantages of each system which must be taken into account.

Viewfinder cameras

The name comes from a separate viewfinder which has disadvantages compared with the SLR viewfinder which 'sees' through the lens. Using the separate viewer, it is difficult to frame a picture and it is therefore more difficult to focus on a particular subject. Focusing at close range is especially problematical because of the reduced depth of field. When photographing close-ups with the macro extension rings fitted, a fixed distance rod and frame must be used. With such a rig fitted to the camera it is naturally not possible to photograph a large subject. Autofocus, macro and zoom lenses cannot be utilised with viewfinder cameras. The advantage of relatively low price also diminishes if purchases of accessories and lenses are made. Then there is the perennial question before a dive into the unknown of which lens to mount. How can you predict what you are going to encounter on a particular dive? The main advantages of a viewfinder camera apart from the low price are

small size, light weight and the fact that a standard lens can be converted to a wide angle by simply putting an additional wide angle lens on top of it.

SLR cameras

These have the advantage that you can frame and focus on your subject without taking a leap into the dark, because the diver-photographer sees exactly what the lens sees. Most camera systems today have an AF (autofocus) installed, which can automatically adjust the lens to focus on a subject at different distances. In theory autofocus is faster than manual focus. The AF measuring system controls the shutter release and a picture cannot be taken when the object is not in focus. This is a great advantage when taking animal pictures in poor light, as is usual underwater.

But not all AF systems are suitable for underwater photography, only those with contrast measuring focusing systems can be used. The use of macro lenses enables the photographer to take pictures of all sizes of subjects from close ups to distant views. For example a small crab of 2 cm can be photographed immediately after photographing a shark

If purchasing an AF camera, before taking it underwater, practise with the system on land without a film in it. Practice in operating camera controls is especially important when taking close-up shots.

Cameras in waterproof housings are larger and heavier than underwater viewfinder cameras, but they can be utilised also on land without a housing.

In addition all kinds of different lenses are available from the macro to the fisheye, and by fitting a zoom lens one can adjust the camera as required. Depending on the design and manufacture of the housing, most are neutral in weight underwater. An experienced photographer appreciates this. In case you decide on a camera housing, you must check which controls are built in. You should consider the following:

autofocus on/off,
zoom operation,
programme change,
+/- correction,
aperture,
shutterspeed.

Single lens reflex underwater camera RS

This expensive camera, even with the standard 28mm lens and flash, is beyond the price range of many divers. Underwater the sturdily-built camera

A flexible underwater housing for an auto-focus SLR camera with flash attached, designed to be used at the relatively shallow depths where the light is best for the inexperienced photographer. (Ewa-Marin)

A special type of lens is needed to take wide-angle photographs underwater, because the magnifying effect reduces the field of view of ordinary wide-angles used in housings. The UW-Nikkor 15mm lens, below, can not be used for photography on dry land and must be immersed to focus correctly.

A compact underwater camera needing no additional housing (Sea & Sea Motormarine)

is not buoyant. It can therefore be uncomfortable to hold over a long period whilst attempting to photograph shy fish. The wide-angle zoom lens (20-35 mm) is limited in scope to seascapes, divers and large gorgonians. Small subjects cannot easily be photographed full-frame. A zoom lens (28-35 mm) would be more useful.

The advantage of the RS is the large viewer which shows the actual settings on the side. This can be rechecked before taking a picture. The hope of many underwater photographers to have a SLR camera the size of the Nikonos was not fulfilled. Especially with the zoom lens, the RS is nearly as large as a camera housing for a land camera.

Underwater flashes are specially manufactured and must be of high quality and sturdy. TTL flash control and a pilot light are a great help for photographing with autofocus (Hartenberger Magnum 250)

Any choice is difficult, so too is the proffering of advice. The best I can do is to introduce you to my favourite camera. My choice was a Minolta 9xi, which has a very fast autofocus. Additionally the AF system has 43 small AF sensors which can be activated when needed and it is possible to use a chip card with this system. The card DATA 2, is a useful learning tool since it records technical data from the previous four films. After a film is developed, the settings of many shots can be recalled to check accuracy. The customised function xi setting on the chip can also be used to re-program many camera functions.

The Flash

As soon as you dive into the water daylight is considerably reduced, because a large number of light rays are reflected from the surface. The greater the angle at which the light strikes the water, the more light is reflected. The best time for daylight photographs is noon, when the sun is at its height.

Light is further reduced with increasing depth because floating particles scatter light. A diffuse light results, low in contrasts. The water also has a strong filter effect, absorbing the colours in the following order: red, orange, yellow, green and blue.

This is why, when photographs are taken underwater without a flash, they are bluish and poor in contrast. Because of the small amount of light remaining, the aperture has to be opened wide: the depth of field, especially at close quarters, is reduced to a minimum. These disadvantages can largely be removed with a flash. Nowadays almost all flashes in use are electronic. TTL ("through the lens") flash technology makes it possible to photograph subjects at different distances without having to adjust the flash. This is very useful with moving subjects, as you never know beforehand how close a fish will come to the camera.

When buying a flash, never buy one of poor quality, because when taking pictures against the light the aperture has to be almost closed. A diver some two metres away will then be so poorly illuminated that he will only appear as a silhouette. A flash that needs to cool off for five minutes after only five full flashes is also insufficient. The manual does not cover explaining to an interesting underwater creature that it will have to wait that long! If you forget to switch off the flash just once, the batteries will be flat within an hour.

The best equipment in my experience is the Magnum 250 from the Hartenberger company. The device is ready for use so quickly that you scarcely have to take that factor into account. The manufacturer guarantees 36 full flashes one immediately after the other. The pilot light, an aid to AF at night, in caves or at greater depths, has two performance levels. The plugs of the synchronising cable are sealed with two O-rings, so that no water can get into the plug threads and no lime scale deposit is possible. If you should forget to switch the flash off the batteries are still not flat even after a couple of hours.

If you take photographs by day you should try to capture the daylight and the impressive underwater landscape with the blue of the water. With a small aperture ($f22$) the depth of field is very great, but daylight will not be enough to illuminate the film. The photograph will then look like a picture taken at night, as the background will be black. In order to achieve good mixed lighting pictures, the daylight should be underexposed by 1-2 apertures, otherwise the picture will seem too pale.

Animal photography

There are may species of marine animals in tropical seas which make easy work of photography. It sometimes seems as if acting as photomodels is their favourite occupation. However, these are almost always the same species: lionfishes, batfishes or the popular anemone fishes, which cannot leave their anemones.

Then there are the territory-bound species, which do not try to escape but turn their backs with an impertinence that could make you tear your hair out, since you never catch them the way you want them. Among these are the radial lionfish, which you rarely see in a photograph from the front or diagonally from the front.

With the fast swimmers that only pass by the reefs occasionally you will have to be quick and make sure the first shot is right, as there often will be no second chance.

Many underwater photographers make the mistake of swimming towards creatures. This achieves the opposite of what they are after. It is the quickest way to frighten away timid creatures, and the others will come up close by themselves. So why swim towards them?

There are also many species which seek protection in caves when approached, such as groupers, sleepers and sentinel gobies. However, groupers have become used to being fed in many places and are therefore easy to photograph. The only way with small species is to exercise patience. If you do not have it, you will not be bringing any good pictures of these species home.

The most difficult to photograph are creatures which can flee over long distances. Sometimes chance takes a helping hand. You can try for ages to get a certain animal close enough to your lens. Then, moving to another area, you find that the same species has quite different behaviour patterns and will be swimming about under your nose. Seasonal changes, such as mating behaviour or caring for the young, will also sometimes bring you long-awaited success. However, one important condition is your own behaviour. Photography of living creatures is often described as a photo-"hunt". Just as a predator targets its prey, we target the subject of our photograph with our lens. The creature sees this as a danger and moves out of the way. The more it is followed, the more timid it becomes. It is better, therefore, to approach an animal as if by chance and observe it only occasionally out of the corner of an eye. This is a quicker route to good successful pictures than trying to swim in pursuit of a fish.

You should also consider beforehand from what angle you would like to photograph a creature. A picture of a fish taken from the side may be a good photo for identification but could be considered dull. However, an impressive photo taken from the front for effect is almost useless for identification. Every kind of photography has its charm, you just need to be clear what you want to achieve.

Film for underwater photography

As in all things, tastes differ widely here. However, it is wise to do without print films, although some photo-graphers have achieved notable successes with them.

If you have prints made from a print film, the colour printing machine will compensate for incorrect exposure and colour. The unusual light qualities underwater are seen by the printer as errors and corrected, even if the film was perfectly exposed. The results are usually unsatisfactory. If you want to avoid disappointment, it is best to move straight on to slide film.

Every company has its own manufacturing process which has a slight effect on the colour repro-duction; Agfa, for instance, favours pastel shades, Fuji tends rather to the opposite. These differences are not important in photographs taken on land, but with underwater photographs the effects are more extreme.

The accepted opinion among most experts is that Kodak slide films give the most natural colour reproduction in this specialised field. There are two different films, Kodakchrome and Ektachrome; both are available with different light sensitivity. Ektachrome has the advantage of taking less time to have developed almost everywhere. This could give you the opportunity of checking your first holiday photos and avoiding any future mistakes.

The Kodakchrome films 25 and 64 have long been top favourites among many underwater photographers, with the 25 being specially suitable for macro photography. You will need a strong flash, so that you can work with a small aperture. The 64 is preferable for mixed light photography, in order for the depth of field to be sufficient.

With a new camera or new flash it can

do no harm to shoot a trial film before going on holiday, as the manufacturer's claims do not always match up with practical experience. You do not need to make an elaborate fresh water diving expedition; a swimming pool with plenty of daylight will be enough. A bathtub would do for macro or close-up photography.

A useful tip to avoid the white sides reflecting too much light is to line the bath with cloth that should not be too light or too dark. Plastic flowers, dolls or other colourful toys can serve as handy subjects for photography. The knowledge gained from the results could have a very positive effect on the next set of holiday photographs.

If films have been wrongly stored the colours may change drastically, and the contrast can suffer as well. It is only heat that damages film: films which are kept cool can be stored for years. Time stops in the fridge! But do take care not to put the film into the camera straight out of the fridge. It needs to adjust to the ambient temperature so that no condensation can form.

Protecting the environment

Looking at the way human beings treat the oceans, one could imagine that they were dealing with a self-regenerating waste disposal site. Toxic substances, acids, oils, plastic and other rubbish are dumped in the sea – out of ignorance, carelessness or selfishness. It is well known that even a small amount of interference with nature can have large-scale consequences, but where the oceans are concerned nobody seems to have many worries about ecological balance. No-one seems to fear a possible damaging reaction, and certainly not that the situation might become life-threatening.

Measured against our human span, the sea has existed for so long that it is almost impossible to comprehend. And yet within a few years human beings have succeeded in creating clearly visible changes along large stretches of coastline. For years scientists and observers of nature have warned about this destruction, but hardly anything has changed. People seem to be trying to outdo one another in construction and progress, which unhappily often means destruction. Even if these damaging influences have only brought change to relatively small areas – when measured against the size of the oceans – we do not know for certain where the limits lie.

The biggest problem of our age – overpopulation – is being ignored by the human race. We are continuing our uninhibited expansion.

All prognoses about population growth on the planet have been exceeded by far. It seems obvious that neither politicians nor religious leaders have any interest in doing anything about this situation which threatens our descendants. Indeed, it would seem that the opposite applies. Population has increased considerably in western European countries over the last 20 years, but some politicians are more concerned that old age pensions are in danger because we are not having enough children!

Overpopulation is steadily increasing the burden on the natural world.

This burden is further increased by the rising standard of living, and the progressive increase in the waste products of "human civilisation".

Some of these products are washed into streams and rivers by rain or are directly poured in by human agency. Apart from natural products such as leaves, sand and mud, toxic chemicals, plastic and fertiliser are taking the same route. Day by day, year by year huge quantities of such matter are being stored in the sea, but not only in the sea; they are also stored in the bodies of fish. When the fish reach our tables as food, the harmful substances, which we ourselves have produced, enter our own bodies.

The amount of poison is not yet great enough to cause panic, but chemical production continues unabated. Many poisons that we produce remain with us, in some form or another.

There are already places on earth today where people do not have enough food. The greater the population density becomes, the more areas will be affected. The numbers of fish in the seas have been noticeably reduced, and it is time now to fish less, to allow the numbers to recover and to preserve this source of food. In practice, however, nets are being made with smaller holes in order to catch as many fish as before. This means the numbers are being reduced further and the food supply of the future is even more in danger.

In the Philippines, fishermen have not been able to meet the demand for food from the sea by conventional methods, so they are now fishing with dynamite. The disastrous effects of explosives underwater are well-known. Only a small proportion of the fish that are killed float to the surface. 90% sink to the bottom! The number of small fishes, larvae and eggs killed is unima-ginably high. Unhappily the coral reefs themselves are so shattered by this process that no fish will settle on the destroyed reef for a few years, as there are no hiding places left. Even stone coral will not grow on coral rubble.

Dynamite fishing is illegal everywhere, but in the Philippines it is not possible to prevent this destruction. At first, the police claimed they could not catch the dynamite fishers because their boats were too fast. This cry for help led to donations from Europe, in the shape of speedboats.

Then, when it became necessary to move against the dynamite fishers, the police shrugged their shoulders: no petrol. There is always some excuse. However, nobody talks about corruption and nepotism, because everyone is trying to fill his own pocket – all at the expense of the natural world.

Even if the coral reefs took a long time to recover, numbers of fish would improve considerably if dynamite fishing was to be stopped. The differences between those who want to protect these unique habitats and those who exploit them ruthlessly are so great that communication is almost impossible, as neither side can understand the other. There is little hope, therefore, that things will change.

The love and devotion with which many aquarists care for their tropical fish and invertebrates in marine aquaria is admirable. They will spare neither expense nor effort where the well-being of their pets is concerned. By contrast, the methods used by local people in catching the creatures in the coral reefs are positively grotesque.

Only a few of the fish caught reach their destination alive and healthy. If only the aquarists realised that they were financing these brutal methods by purchasing the fish, many would surely do without.

The methods of catching fish vary greatly. The fishermen do not regard a fish to be caught in the same way as an aquarist, whose aim is to get the creature, with a lot of patience and care and undamaged if possible, into an aquarium. Fish are merely a source of income, on which they waste no emotion. Accordingly, they are ruthlessly treated.

Netting fish is one of the more harmless methods: the fishermen use a lot of lead and stamp about on the bottom around the fish they are trying to catch, breaking off all the coral. The fish is packed into a plastic bag and put in the boat, where it often lies in full sunlight. The water temperature rises and oxygen becomes scarce. Add the stress of the catch, and soon you have the first fishes lifeless in the bag. Once the fishermen realise that the fish cannot be preserved alive for the aquarium, it is of no more value to them. It is thrown overboard, sometimes still in its plastic bag, for plastic in the Third World is cheap. But even if a fisherman takes the trouble to throw the fish back in the sea without the bag, the fish has almost no chance of survival.

There are, however, much worse methods. The greatest damage is done to the reefs by poison which is sprayed among the corals to make the fishes leave their hiding places. Only a very small proportion of the fish caught will survive. Even more affected, though, are the sessile life forms, such as corals, which cannot escape from the cloud of poison. In some countries (such as Indonesia, Sri Lanka and the Philippines) large parts of the reefs are being destroyed in this fashion. This is surely not the intention of marine aquarists. But it is hard to do without fish for the aquarium. Doing without, however, is the only way in which this dirty treatment of the natural world can be stopped.

There are of course many other topics which cannot be stressed often enough, but from experience we know that there is much talk about protecting the environment but far too little is being done. This is why I want to confine myself to points which divers can keep in mind if they have a proper awareness of the environment.

Unfortunately, many people believe that their "minor faults" do not count when set against the destruction being carried out today. However, if everyone thinks this way, those who cause damage on a large scale can find excuses for their behaviour. Finally, the total amount of many little mistakes by a large numbers of divers can put quite a strain on some environments.

The right way to act

In certain places used by divers, for instance off Sinai, there is scarcely a live coral to be found around the places where the divers enter the water. This is not so much because the divers are consciously ruthless in their actions. Most feel close to nature and want to see living creatures in their natural surroundings – that is why they learned to dive in the first place. You can often meet beginners who are astonished when they discover that corals are living organisms. The damage is therefore mainly due to the fact that very few diving schools impart such important information. And how is a new diver, seeing this new world of the tropical seas for the first time, to know what it is that he sees?

It is not my purpose here to hand out blame for actions done in the past, but to give some advice on ways of keeping this splendid natural environment as unchanged as possible. What diver would not like to find, after several years' absence, that a fondly remembered diving spot was as beautiful as ever? Even small, trivial actions to which almost no-one pays attention can damage corals.

Almost all divers kick up sediment with their fins – some more than others. Among the most important vital functions of the coral polyps are catching food and keeping the coral clean. If a coral is covered in sand it must immediately be cleaned by the polyps. If the burden of sand in some areas is great, the polyps are so occupied in cleaning the coral that they have hardly any time to catch food, and the colony dies.

Of course sediment is also raised by sea creatures, by goatfishes looking for food, for instance. However, this happens at such long intervals that the corals are not endangered.

Buoyancy is the most important diving skill which a diver should have. It is possible to pass a diving course in a relatively short time, but diving, just like driving, is properly learned only by practice. Students and beginners should not dive straight into a coral reef, because they often touch the bottom and break off coral without meaning to. It would be desirable for diving instructors not to be so keen to show beginners as much marine life as possible, but to put more stress on the following:

- lead weights should be steadily reduced as much as possible:
- swimming position should be improved;
- learners should remain calm when in involuntary contact with the bottom and not strike out with their flippers
- learners should raise themselves correctly from the bottom without breaking off coral or whirling up sediment.

When descending, the specific gravity of a diver will increase due to the compression of the air in the equipment. If you want to get down deep quickly, you will find this useful. However, before you reach the bottom, you will need to get enough air into your buoyancy jacket so as not to hit the bottom hard. Avoid touching the bottom as much as possible!

A large number of divers use more lead than is necessary. To counterbalance this, more air is needed in the buoyancy jacket. This means that the body is not level; the legs sink downwards. In this diagonal position it is more likely that coral will be broken and sediment kicked up. Use as little lead as possible!

Divers use many opportunities for holding on under water. Many also wear protective gloves because they are worried about injuries. However, this makes them grab harder and much more carelessly, and far more damage is caused. Good divers do not need protective gloves!

Many snorkellers and divers going from the shore into the water walk as far as they can in shallow water in order to avoid swimming. But even in the shallow waters of tropical seas there are many corals and other delicate organisms. Do not step on corals!

Many manufacturers of flippers are more concerned with design than with practicality. For many years flippers with so-called JET-slits have been manufactured and sold. However, these slits have the disadvantages of making the diver more likely to get stuck on coral. Smooth flippers, with a whip-like effect, are much better and more effective. Do not use slitted flippers!

More and more often individual divers can be seen swimming through the reefs with knife in hand as if they had to defend themselves against some enemy. Unfortunately, the knife is used to break up sea urchins or to feed mussels to fishes. This is interference with the natural ecological balance of the reefs and as damaging as collecting molluscs for their shells. Harpooning is of the same order. This topic has been discussed over and over and should by now be resolved, but the harpoon freaks are still out there. It has been common knowledge for a long time that this practice not only decimates the fish populations but also makes all fishes timid, so that they can no longer be observed at close quarters. Do not kill any creatures! (Angling does hardly any damage to the reefs.)

Buying dead shells of molluscs apparently does no damage to the sea, but this appearance is a delusion. Most creatures are only killed if there is a demand. If no-one was to buy the shells, the animals would have at least a much better chance of survival.

The killing of vertebrates for the same purpose is even more ruthless. Sea horses are caught, dried and sold to tourists. Our selfish desire to possess everything that pleases us is fatal for many creatures. Porcupine fishes, for instance, are useless as food but they are filled up with sand, while still alive, and dried. They are then sold as lampshades or ornaments. Fan corals with many polyps in them are dried for the same purpose. Amazingly, those who buy such souvenirs still believe they are innocent!

Every kind of collecting of living creatures damages the natural world.

Collectors pay high prices for the triton's trumpet, a large marine gastropod; as a result, the numbers of crown of thorns starfish have increased drastically in many coral reefs, as their natural enemy, the triton's trumpet, has been decimated. The crown of thorns starfishes eat coral polyps. Only a few species of marine life can survive in dead reefs. The best course is not to buy any souvenirs from the sea!

Feeding creatures in the sea is very popular and is practised in many places. Coral fishes, morays and even sharks are fed by hand and many become tame. I must admit that I once used to do this myself because no negative experiences had been recorded. The disadvantage is that many divers see the feeding and it encourages them to imitate such experiments. Enough accidental bites, with some serious injuries, have resulted.

In many places in Egypt I noticed that the large morays were becoming quite importunate, following divers several metres out into the open water. It is quite possible that unprepared divers would be startled, react in the wrong way and then get bitten. But even feeding the coral fishes has its disadvantages; shoals of various species often swim so close to the divers that there is no opportunity to observe their natural way of life. It is better not to feed the creatures in the sea!

Divers have been catching animals for a long time. They let themselves be pulled through the water by turtles; they catch puffer and porcupine fishes to make them blow themselves up, and hang onto harmless nurse sharks for fun. Diving magazines are not innocent of encouraging these developments: such action photographs were being published only a few years ago.

However, this is no fun for the creatures concerned, but a situation of extreme stress, as they do not know that the divers are not really enemies. Puffer and porcupine fishes can die of this stress. Nurse sharks which have been touched by divers do not react spontaneously if the contact is not too hard, but they will never return to that particular spot. In this way interesting spots for divers can be ruined in a short space of time. Please do not touch any creatures in the water!

Filming and taking photographs is a very fine hobby and gives many who cannot see the coral reefs for themselves an insight into the variety of this habitat. However, it is unfortunately not possible to see from the pictures how they were taken. Many photographers forget that there are living beings outside the viewfinders of their cameras: they wind their legs around corals to hold themselves in place as if they were in a gym. Wrapped in neoprene, they squash themselves ruthlessly into cracks which are full of living organisms, just so that they can take their photos in peace even if there is a current. Please take more care of invertebrates!

When diving at night, lights are necessary. There is a wide range of lamps available. Relatively strong lights are needed for filming to make use of the total range of splendid colours of the underwater world. These lights, though, are often used for night diving, although there is no need for them. Many of the resting or sleeping creatures are woken and stagger sleepily out of their usual sleeping places, which they later cannot find again. They are then in danger, as they cannot find their way back to their hiding places until the next day. Ten watt lights are sufficient for night diving!

Diving and boats go together – they are almost inseparable. Diving without boats is only possible in a few areas. Anchoring can largely be avoided if in frequently-used places there is a "harbour anchor" fastened to the bottom and equipped with a buoy to which the boat can be fastened. The boat can stay in the vicinity during the dive without having to drop the anchor into the delicate corals.

With cruises, the rubbish problem is an additional factor. The sea can easily deal with all organic kitchen waste. However, plastic, cans and batteries impose a heavy strain. If there was space enough to take it when the boat set out, it is reasonable to expect that such rubbish should be taken back again.

In the evenings in good company many a beer can closure is thoughtlessly flung overboard. And there are enough people who fling not just the closure, but the whole can overboard without thinking anything of it! Do not throw any non-biodegradable waste into the sea!

For many years large quantities of consumer goods have been imported into tropical diving areas. For example in the Maldives, for tourism purposes. Drinks are needed en masse. Although it is well known that glass is not a serious strain on the marine environment most drinks, even after many years of experience, are still delivered in aluminium cans.

Unfortunately only a few diving schools and clubs are teaching their students basic knowledge about the protection of the seas. However, it is one of the instructor's most important tasks to pass this knowledge on to the students, so that every diver can avoid thoughtless errors right from the start.

Further reading

Allen, Gerald R. & Steene, Roger C.:
    Reef Fishes of the Indian Ocean.
    T.F.H. Publications 1987.
Carcasson, Robert H.: Coral Reef
    Fishes.
    William Collins Sons & Co.
    Glasgow 1977.
Cuthbertson, Lydia: Guide to the
    Maldives
    Immel Publishing
    London, 1994
Myers, Robert F.: Micronesian Reef
    Fishes.
    Published by Coral Graphics,
    Guam, 1989.
Randall, John E.: Diver's Guide to
    Fishes of the Maldives
    Immel Publishing
    London, 1983
Randall, John E.: Red Sea Reef Fishes
    Immel Publishing
    London, 1983
Vine, Peter: Red Sea Invertebrates.
    Immel Publishing, London, 1986.
White, Alan: Philippine Coral Reefs.
    New Day Publishers, Quezon City,
    1987.

Glossary of specialist terms

Adhesive capsules little 'blisters' in cnidarians which burst when touched and shoot out an adhesive thread which holds prey

Benthos animals or plants living on the bottom of seas

Ciguatera poisoning due to ciguatera toxin, which occurs in certain algae (dinoflagellates) which are eaten by herbivores. The greatest concentration is found in predatory fishes at the end of the food chain.

Cilia (biological:) very fine hairs which beat all in one direction

Cirri or tube feet: tentacle-like body attachments of animals which have a propulsive function

Claspers cone-like structures on the anal fin of sharks, one of which is inserted into the female's cloacal opening during copulation

Cloacal opening anal opening which also leads into the sexual organs

Cuvian threads gland-like attachment at the end of the intestine of many species of sea cucumber

Denticles scales of catilaginous fishes

Detritus decaying tissue parts of animals and plants

Endemic limited to a certain area

Grazers animals which feed on the plant or animal growth covering the bottom

Hermaphrodites organisms with two sexes which can produce both sperm and eggs, either both together at one stage of life or consecutively by changing sex (very common in fish)

Hermatypical reef-building

Larvae creatures at the young stage of their development, often looking unlike their parents; the organs may also not be fully formed

Medusae jellyfish: a pelagic generation which reproduces sexually

Nekton animals living in the water which can propel themselves forward against the current

Operculum 1. cover for closing gastropod shells; 2. gill cover in fishes

Oviparous laying eggs before fertilisation or at an early stage

Oviviparous laying eggs: the embryos are developed and ready to hatch when the eggs are laid

Pelagic living in open water

Phytoplankton plant plankton

Polyp sessile invertebrate with a sack-shaped body and a mouth surrounded by tentacles

Sediment matter laid down on the sea bed

Sessile attached to one place

Solitary living as a single individual

Spermatophore complicated structure of stuck-together sperm cells

Stinging capsules (nematocysts) little 'blisters' on cnidarians which burst on contact and inject a tube into the skin of an organism through which the stinging poison flows. Used for defence or to catch prey

Tetrodontoxin powerful poison found in the inner organs and skin of pufferfishes

Viviparous bearing live young

Water-vascular system fluid-filled tube system of echinoderms

Zooplankton animal plankton

Zooxanthellae algae which are embedded in the tissues of animals and live in symbiosis with them

Index